**Labour Administration
Training Material**

LABOUR INSPECTION AND ITS ROLE IN IMPROVING SAFETY AND HEALTH, WITH PARTICULAR REFERENCE TO THE CHEMICAL INDUSTRY

TM. 8

Labour Administration

TRAINING MATERIAL

LABOUR INSPECTION AND ITS ROLE IN IMPROVING SAFETY AND HEALTH, WITH PARTICULAR REFERENCE TO THE CHEMICAL INDUSTRY

Training Material and Report on
National Seminars in Hong Kong
(8-12 and 16-20 June 1987)
Organised by ILO/ARPLA
on Labour Inspection and its Role in Improving Safety
and Health, With Particular Reference to the Chemical Industry

International Labour Organisation
Asian and Pacific Regional Centre for Labour
Administration (ARPLA)
Bangkok

ISBN 92-2-106297-X
First published November 1987

Printed in Thailand

PREFACE

In the High-Level Meeting of the ARPLA Tripartite Technical Advisory Committee held in August 1985 and the Tripartite Project Review Meeting held in December 1986, Government representatives identified a number of emerging problems which directly or indirectly impinge on labour administrations. It was felt that within the limits of the resources likely to be available to them in the future, ARPLA should endeavour to tailor its activities to take account of these emerging trends and their implications.

One of the areas of growing importance identified concerned the proliferation of hazardous processes and substances in industry arising out of technological development. Labour administrations are increasingly being confronted with various types of hazardous processes and substances, which are placing further pressures on labour inspection services. In order to enable labour inspection services to cope with the situation, ARPLA has been making efforts to equip them with the inspection skills needed for monitoring and control of occupational hazards arising out of the transfer of technology, particularly due to the growing use of chemicals. An Inter-Country Training Course on Inspection Skills in the Chemical Manufacturing Industry was organised by ARPLA in Perth in 1986. This course was found very useful and was repeated in Perth in 1987 at the request of the participating countries.

The Hong Kong Government's Labour Department further requested ARPLA to organise two national seminars in June 1987 on Labour Inspection and its Role in Improvement of Safety and Health, particularly covering inspection skills in industries using chemicals. We have produced a composite training handbook containing the technical material used in the Hong Kong seminars which we hope will be found useful by labour inspectorates, employers and workers in the Asian and Pacific region.

We hope that the training volume will bring about a greater awareness of the importance of labour inspection and its role in the improvement of safety and health, with particular reference to the chemical industry.

A.M.A.H. Siddiqui
Chief Technical Adviser

Bangkok
November 1987

CONTENTS

Page

PREFACE . . . . (i)

REPORT OF SEMINARS 1

    General Aims and Objectives 3

    Specific Objectives 4

    Structure of Programme 4

    Contributors to Programme 5

    Seminar Evaluation 5

    Conclusions and Recommendations 5

SELECTED TECHNICAL PAPERS 7

    Conceptual Model of Accident Process, System Safety and Productivity 9

    The Concept of Accident Cause 17

    Bhopal – The Worst Occupational Disaster in History 19

    Physical Hazards: Fires and Explosions 27

    Prevention of Dust Explosions 47

    Recent Developments in Major Hazard Control Methods 57

    Techniques of Inspection of Major Hazard Works 67

    Release of Flammable and Toxic Substances and Effects of Fires 77

    Explosions 85

    Health Hazards from Chemicals in the Workplace 91

    Disaster Planning and Chemical Accidents 123

    Legislation and Exposure Standards for Chemical Substances 135

    Sampling Strategies and Techniques for Contaminant Evaluation 143

    Industrial Hygiene Audits 147

    Personal Protective Equipment 161

    Hazardous Materials Management 167

    Disposal of Hazardous Wastes 177

ANNEXURE 185

    I. Syllabus of Seminar Programme 187

REPORT OF SEMINARS

REPORT OF SEMINARS

The number of occupational accidents registered a sharp rise in Hong Kong in 1986. Adding to the heavy accident toll in the construction industry, the chemical industry and factories involving the use of large quantities of hazardous chemical substances in their processing contributed substantially to the total industrial accident toll. In October 1986, there was an explosion and fire accident (due to the use of highly flammable solvents in a fur dyeing factory) which claimed 13 lives, in addition to causing injuries to several others.

To strengthen their efforts to ensure safety at work, the Government of Hong Kong made a special request to ARPLA to organise national-level seminars, on an urgent basis, on Labour Inspection and its Role in Improvement of Safety and Health.

As the Hong Kong authorities wished that all their inspectors and other officials engaged in inspection should participate in the training, it was felt worthwhile to have two seminars. The first was held from 8-12 June 1987, and the second, from 16-20 June 1987, was a repeat seminar. Two hundred Inspectors from the Factory Inspectorate were given training. The Government of Hong Kong also invited 7 Inspectors from Macao to participate.

GENERAL AIMS AND OBJECTIVES

The general objectives of the seminars were:

(a)   to provide a basic conceptual framework for the study of occupational safety and health;

(b)   to bring together these disciplines within the conceptual framework, in order that the principles of occupational safety and health can be applied in practical situations defined in relation to the work and experience of the participants. In particular:

    (i)   to develop and expand the participants' knowledge in their own fields of expertise, with particular emphasis on using field testing techniques and equipment; and

    (ii)   to assist participants to adjust themselves to new working procedures and conditions and to develop further inspection skills when hazardous materials and processes are involved.

SPECIFIC OBJECTIVES

The specific objectives of the seminars were to enable participants:

(a)   to carry out a health hazard survey of a particular process and suggest the methods of minimising possible loss and damage;

(b)   to assess the fire and explosion risks of certain materials;

(c)   to propose safe methods of storage, handling and disposal of hazardous materials;

(d)   to assess, quantitatively, the safety of a process;

(e)   to suggest an efficient protective system for certain chemical operations and industries;

(f)   to understand the scientific, circumstantial and organisational background to fire and explosion;

(g)   to explain the role and functions of labour inspection, with reference to working conditions, and occupational safety and health; and

(h)   to advise on organisational and procedural aspects of labour inspection with regard to the implementation of national standards.

STRUCTURE OF PROGRAMME

The seminars were conducted in accordance with a carefully selected programme drawn up by ARPLA, in consultation with the Chief Factory Inspector, Hong Kong, and three resource persons from Australia.

The programme comprised lectures and tutorials, averaging approximately thirty-five hours per week and covering the syllabus shown in Annexure I. In addition, participants were asked to undertake individual study.

The programme commenced with presentations on latest concepts of accident prevention, modern approaches to chemical hazard control and an introduction to health hazards and industrial hygiene. This was followed by presentations on inspection skills for the prevention of chemical accidents, on the identification and control of chemical health hazards (specific to occupational health hazards in Hong Kong), and on the exposure standards for chemical substances. Other presentations related to topics like the disposal of toxic wastes; assessment of the consequences of fire, explosion and toxic releases; Dow fire and explosion index technique and hazards and operability studies. The biological effects of various common chemicals were discussed in detail. Keeping in view the nature of the industry in Hong Kong, a presentation was made by ARPLA on low-cost means of controlling chemical hazards in small enterprises. The presentations are given under the section "Selected Technical Papers".

4

CONTRIBUTORS TO PROGRAMME

As in all seminars of this nature, the wide range of interests and backgrounds of the participants required contributions from widely experienced multidisciplinarians and interdisciplinarians.

ARPLA Representative

Expert in Labour Inspection.

Programme Director

Dr. M. Nedved, Senior Lecturer in Occupational Safety and Health, School of Community Health, Curtin University of Technology.

Expert Lecturers and Seminar Leaders

Dr. Peter Hollingworth, Consulting Occupational Physician; part-time lecturer, School of Community Health, Curtin University of Technology.

Mr. Barry Chesson, Chief Occupational Hygienist of Alcoa of Australia; part-time lecturer, School of Community Health, Curtin University of Technology.

SEMINAR EVALUATION

Participants were each requested to complete a questionnaire prepared by ARPLA in collaboration with the West Australian team. In response to the questionnaire asking for an overall assessment of the seminars, as a whole, on a 5-point scale (ranging from 'of very little' to 'of very much' value), 166 of the 200 participants felt the seminars were of real value. One hundred and seventy participants, in a similar rating, indicated that they had increased their knowledge, and 154, that they had added to their skill levels for effective labour inspection. The lectures and tutorials were appreciated by about 180 participants. The topics were found very relevant by the participants.

CONCLUSIONS AND RECOMMENDATIONS

The participants appreciated the training and its relevance to their needs. Some changes were made in the lectures for the second seminar on the basis of comments made by the participants at the first seminar. The participants not only learned about latest developments in labour inspection but also benefited from their interaction with experts and practitioners in the field. The knowledge gained by the participants would enhance their skills in inspecting industrial units using common chemicals and in recommending methods for minimising possible loss and damage.

For future programmes of this kind or of a similar nature, it is recommended that ARPLA distribute in advance preliminary reading material, such as the ones developed for these seminars. This would enable the participants to benefit more from highly demanding and condensed programmes in the field of occupational safety and health.

Hong Kong is keen that such activities should occupy a prominent place in ARPLA's future programme. The Chief Factory Inspector, Hong Kong, has constituted a committee of senior officers for a detailed evaluation of the contents of the presentation papers made at the seminars. The results of the assessment will be discussed with ARPLA's experts at a future date with a view to drawing up the content of a possible follow-up seminar.

SELECTED TECHNICAL PAPERS

CONCEPTUAL MODEL OF ACCIDENT PROCESS, SYSTEM SAFETY AND PRODUCTIVITY

by
Milos Nedved

(Paper Presented at the International Conference on High Technology —
Human Error Disasters, held in Perth, Western Australia, 1-3 December
1986)

DEVELOPMENT OF ACCIDENT PREVENTION

Already in 1919, Greenwood and Woods, researchers for the British Industrial
Health Board published their statistical analysis of injuries in a munition factory.
During 1926, Newbold followed this up with a study of 13 factories, which confirmed
the earlier results. These and various other studies at that time explored the relationship between accident rates and temperature, humidity, work hours, age, experience
and absenteeism.

Fifteen years later, applied psychologists asked specific questions about the
efficiency of human production, accidents being a significant element of this efficiency. By that time it was well-recognised that the multitude of factors contributing
to accidents were not amenable to simple and direct manipulation and control.
Parallel to this "pure" accident research, "applied" prevention techniques have been
developed by safety engineers, medical officers and inspectors. The vast majority
of them have resulted, not from accident evaluation research, but from an experienced
understanding of the immediate problems (machinery guarding is a good example — if
the danger and the potential victim are physically separated, no accident can occur).
These techniques have been later improved by applying the modern psycho-physiologi-
cal knowledge of human and environmental interactions.

A number of accident models have been developed during the last 30 years.
Most of them can give very useful insights into the accident sequence — so that such
insights could be used in designing successful accident prevention strategies. The
oldest, and perhaps the best known model — the Domino model — was developed by
Heinrich. The Domino theory explains the accident process in terms of five factors:

(a)   ancestry and social environment;

(b)   fault by the person;

(c)   unsafe act (procedure) and/or physical hazard;

(d)   accident; and

(e)   injury.

These factors are of a fixed and logical order. Each one is dependent on the one immediately preceding it, so that if one is absent, no injury can occur, which can be visualised as 5 standing dominoes when subjected to a disturbing force. When the first falls, the other 4 automatically follow, unless one of the factors have been corrected, i.e. removed, thereby creating a gap in the required sequence for producing an accident.

The more comprehensive model, putting much emphasis on human factors, was developed by Surry, and is entitled: A Decision Model of the Accident Process. According to Surry:

the predisposing characteristics:

(a)   man (person) susceptible to hazard;

(b)   hazardous environment;

(c)   injury-producing agent,

and

the situational characteristics:

(a)   risk-taking;

(b)   appraisal of hazard;

(c)   margin of error;

have a direct bearing on the outcome.

By his action, or even non-action, the danger to a man grows out of the predisposing characteristics. If there are any negative responses to the following questions during the *danger build-up* cycle, the danger will become imminent:

(a)   Is there any warning of the growth of the danger?

(b)   Can the man see this warning (or detect it by any other means)?

(c)   Does he understand the meaning of this warning?

(d)   Does he know how to avoid such potential danger?

(e)   Does he choose to avoid it?

(f)   Is he able to avoid it?

If the replies are all positive, the danger will not grow and no injury can ensue.

The same series of questions may be formulated in the *danger release* cycle:

(a)   Is there any indication of the danger release?

10

(b)   Does the potential victim detect this information?

(c)   Does he recognise the meaning of this information?

(d)   Does he choose to avoid it?

(e)   Does he know how to avoid the released danger?

(f)   Is he physically capable of avoiding it?

Any negative response to one of these questions will inevitably lead to injury.

SEQUENCE IN ACCIDENT CAUSATION, ACCIDENT INVESTIGATION AND THE CONCEPT OF ACCIDENT CAUSE.

Accidents are usually multifactoral and develop through a relatively lengthy sequence of changes and errors.

The complexity of events leading up to an accident implies that there are many opportunities to intervene or interrupt the sequence. It is, therefore, essential that the *accident investigation process* gives clear visibility of the *complex realities,* not the simplistic categorisation of conditions and acts so often found in accident reports.

Schulzinger described the accident sequence as "a dynamic, variable constellation of signs, symptoms and circumstances which together determine or influence the occurrence of an accident". The use of the word "cause" usually implies that the latter event is the inevitable consequence of the former, i.e. the former event is the direct explanation of the latter phenomenon.

In the accident prevention field, a different meaning applies to merely indicate that there is an association between two events. The safety profession should aim at avoiding the term "cause", since it is too specific and does not admit to the complexity of interrelationship among the multitude of antecedent variables. The use of the term to indicate the mechanism of the accident (e.g. blast, exposure to . . ., etc.) is misleading. Such a "cause" does nothing to explain the factors leading to the accident, nor to indicate a significantly frequent relationship (Surry).

The factors discussed in relation to accidents are not necessarily explanations. They are simply related through some mechanism, or even by chance, e.g. a well-known relationship is that of age and accident rates, but in no way can youthful age cause accidents. The possible link is that of inexperience, which results in performance errors. But even this is hardly a "cause", since inexperience does not always result in accidents. The preferable term should be "contributory factors", instead of "cause", and the assumption of a "solution" by identifying a "cause" must be avoided.

SYSTEM SAFETY

The accident sequence models discussed here, and a number of other models, are very useful in most workplaces and for most processes and operations in the field of classical industrial safety. However, for high technology, and very complex processes and operations, in order to cope with the high number of variables, a more systematic approach is needed. This new approach — system safety — was developed about 15 years ago by American engineers, mainly for the purpose of the space programme. Very soon afterwards, it has been introduced and widely used in some other fields, mainly in the chemical industry. Nowadays, the system safety approach provides a unifying concept for the multidisciplinary field of occupational safety and health.

Industrial safety has existed for several decades, but there is something new about system safety, since it has the characteristics of the systems approach. It is:

(a)   methodical;

(b)   objective;

(c)   quantitative;

(d)   analytical;

(e)   closed loop; and

(f)   interdependent.

In addition, it translates the traditional moral arguments of safety into specific tasks, which can be expressed in the language of top management, such as:

(a)   cost; and

(b)   schedule and performance.

System safety is defined as "the optimum degree of hazard elimination and/or control within the constraints of operational effectiveness, time and cost, attained through the specific application of management, scientific and engineering principles throughout all phases of a system life cycle".

System Characteristics

(a)   *Methodical:* The systems approach involves a definite method. The method consists of an orderly procedure or way of solving complex problems. All the steps involved in problem-solving are arranged in a consistent and orderly manner.

(b)   *Objective:* The systems approach is objective, i.e. the steps in the problem-solving method are free from personal bias, to the greatest extent possible. The results of each step in the problem-solving process can be verified by someone other than the person who performs the step.

(c)   *Quantitative or Measurable:*   Each element in the problem-solving process results in a quantitative expression.

(d)   *Analytical:* The systems approach employs a rational division of the whole system into its constituent parts in order to find out the nature, proportion, function and interrelationship of these parts as they contribute to system objectives.

(e)   *Inputs and Outputs in Clear Language:* Both inputs and outputs, at all levels in the system, are described in clear language, with quantitative indices. In simplest terms, a "system" can be defined as "any complete entity, consisting of hardware, software, personnel, data, services and facilities, which transforms known inputs into desired outputs".

(f)   *Self-Contained/Closed Loop:* A system has been defined as a complete entity. Outputs of every element or sub-system must become part of the system output rather than independent of it. This is a restatement of the fact that everything within the system is interdependent.

System Management

System management and system safety both reach their goals by "accomplishing through others". Therefore, they use the same methods and techniques and must understand one another.

In the past, *industrial safety often ignored management as a source of hazard.* Yet, policies, procedures, design decisions and acceptance of risks lie in the domain of system management. Understanding the function of system management permits system safety personnel to take a total view of hazard control.

Analytical Requirements

Since analysis forms the backbone of system safety activity, and the ultimate purpose of analysis is to aid the decision-making process, there are several requirements for system safety analysis:

(a)   knowledge of the system and its operation;

(b)   knowledge of system environment (natural environment, such as temperature, vibration, etc.; political environment, which is established both by top management and outside forces, like the marketplace; social environment, consisting of ethics and other behavioural standards; economic environment providing forces on the system, which can lead to hazardous operations);

(c)   knowledge of system hazards; and

(d)   analytical methodology (such as FMA, FTA).

Focus for Management Decision

Three parameters of any given hazard are simultaneously considered, and then combined into an overall quantitative index of significance. The parameters are:

(a)   consequence or severity of the particular hazard;

(b)   probability of the hazard occurring; and

(c)   amount and type of resources required to eliminate or control the hazard.

Management can be then presented with a list of hazards according to the degree of significance of individual hazards.

Resolution of Hazards

Deciding whether to eliminate hazards or to make the system tolerant to them.

In summary, system safety is better, when compared with traditional industrial safety, in that it provides a complete and integrated approach for reaching a management decision on any possible hazards within a system.

CONCLUSIONS

Accidents at work happen in diverse circumstances. Each accident involves a combination of many factors and/or the simultaneous occurrence of:

(a)   environmental hazards in the workplace (physical, chemical and biological); and

(b)   human factors (individual and organisational behaviour).

Usually, more than one of these causes contribute to the accident, and with this joint causation, removal of only one type of cause often prevents injury, even though the other causes remain.

Dealing with physical workplace hazards is usually referred to as an engineering approach to safety — trying to "engineer out the hazards". Such a strategy is usually a very successful one and has to be always used in a high technology working environment context.

But, in spite of high technology, accidents arising from relatively mundane situations and activities are responsible for the vast majority of occupational accidents, even in the high technology working environment. The most frequently occurring causes are as follows:

(a)   incorrect manual handling;

(b)   slipping, tripping and falling;

14

(c)   persons striking against objects; and

(d)   objects falling when being handled.

Fortunately, to prevent accidents resulting from the above causes, an expensive engineering approach to safety (designing out the hazards), which is very effective but very costly, is not the only means available to us. Successful interventions in the field of human factors are:

(a)   designing and implementing safe working procedures, including proper supervision; and

(b)   increasing safety awareness, and introducing safety training in order to enable the workers to develop safe working habits — being much more cost efficient — would substantially minimise the frequency of occurrence of such accidents.

I would like to express my belief in the hope that improvement of working conditions and environment would not only reduce the amount of human suffering, but would also be accompanied by an increase in productivity.

REFERENCES:

1.   Greenwood, M., H.M., 1919. The Incidence of Industrial Accidents Upon Individuals, With Special Reference to Multiple Accidents. (Report No. 4, Industrial Fatigue Research Board, Medical Research Committee, Great Britain).

2.   Hakkinen, S., 1978. Accident Theories and Their Development. Laboratories of Industrial Economics and Industrial Psychology, Report No. 36, Helsinki.

3.   Surry, J., 1974. Industrial Accident Research. Labour Safety Council, Toronto, Ontario.

4.   System Safety Handbook, 1971. Naval Ship Systems Command, Department of the Navy, Washington, D.C. 20360.

5.   Thorndike, R.L., 1951. The Human Factor in Accidents, With Special Reference to Aircraft Accidents. USAF School of Aviation Medicine. Project No. 21-30-001.

6.   Heinrich, H.W., 1959. Industrial Accident Prevention. McGraw Hill, New York.

7.   Johnson, W.G., 1975. MORT — the Management Oversight and Risk Tree. Journal of Safety Research 7, 4-15.

8.   Grose, V.L.: Anwendungen des Systemsicherheits — Konzepts der Luft-und Raumfahrt in nicht der Luft-und Raumfahrt Angehorenden Bereichen. Ifu Seminar, Systemsicherheit, Bonn, 1981.

THE CONCEPT OF ACCIDENT CAUSE

by
Milos Nedved

The use of the word "cause" implies that the latter event is the inevitable consequence of the former, i.e. the former event is the direct explanation of the latter phenomenon.

In the safety field, it has a different meaning to merely indicate that there is an association between two events.

Nowadays, there is a tendency to avoid the term "cause", since it is too specific and does not admit to the complexity of interrelationship among the multitude of antecedent variables. The use of the term to indicate the mechanism of the accident (e.g. blast, exposure to . . . , etc.) is misleading. Such a "cause" does nothing to explain the factors leading to the accident, or to indicate a significantly frequent relationship.

The factors discussed in relation to accidents are not necessarily explanations: they are simply related through some mechanism, or even by chance, e.g. a well-known relationship is that of age and accident rates, but in no way can youthful age cause accidents. The possible link is that of inexperience, which results in performance errors. But even this is hardly a "cause", since inexperience does not always result in accidents.

Preferable term: "contributory factors", instead of "cause". The assumption of "solution" by identifying a "cause" must be avoided.

CASE STUDY

A person falls off a ladder.

Classical industrial safety approach:

Unsafe Act: climbing a defective ladder.

Unsafe Condition: a defective ladder.

Accident Prevention Measure: getting rid of the defective ladder.

LOSS CONTROL (SYSTEM SAFETY) APPROACH

To determine some of the contributory factors by asking:

(a)   Why was the defective ladder not found during normal inspections?

(b)   Why did the supervisor allow its use?

(c)   Did the injured employee not know it should not be used?

(d)   Was the employee properly trained?

(e)   Was the employee reminded not to use the ladder?

(f)   Did the supervisor examine the ladder first?

The answers to these questions may then lead to the following accident prevention measures:

(a)   an improved inspection procedure;

(b)   improved training;

(c)   better definition of responsibilities; and

(d)   pre-job planning by supervisors.

By such an approach, we find the fundamental root causes (contributory factors), with the subsequent removal, leading to prevention of a recurrence.

BHOPAL — THE WORST OCCUPATIONAL DISASTER IN HISTORY

by
Milos Nedved

(Keynote Paper Presented at the International Conference on High Technology — Human Error Disasters, held in Perth, Western Australia, 1-3 December 1986)

SUMMARY

Early on 3 December 1984, the safety valve on a storage tank opened as a result of chemical reaction in the tank. At the time, the tank contained approximately 90,000 pounds of stored methyl isocyanate (MIC). It is believed that the safety valve remained open for approximately two hours before it was reseated, and during that period an excess of 50,000 pounds of MIC in vapour and liquid form were discharged through the safety valve, escaped to the surrounding atmosphere and affected a significant proportion of the Bhopal population.

Given the nature of the gas leak at Bhopal, and considering that a substantial section of the population was also exposed to non-lethal doses, the effects of which could manifest over a long period of time, it is very difficult to arrive at a definite estimate of the number of casualties. The Indian Government originally estimated that approximately 1,700 people died. Most of the more reliable news accounts put the figure at around 2,600. The true number will never be known because, in the chaotic situation during the first days after the accident, bodies were buried and burned, without proper identification or even count. But it is likely that both the above-mentioned numbers are too small. There are still thousands of people registered as missing.

The Bhopal hospitals treated at least 130,000 patients for problems — mainly of the eyes and lungs. In addition, over 40,000 patients were treated in 22 other districts of the State. These were, to a large extent, people who fled from Bhopal by whatever means they had. At least 12,000 of the 170,000 patients were in very critical condition when they were brought to the hospitals and wards.

Based on the information available to us at present, the following conclusions about the causes of the Bhopal tragedy have been reached:

The disaster was caused by insufficient attention to safety in the process operations, unsafe operating procedures, lack of proper maintenance resulting in faulty equipment, cuts in manning levels, inadequate training, and lack of skilled supervision. Smaller releases of toxic chemicals had occurred in the past, leading to one death and numerous injuries. Little was done to correct these problems, despite protests by the Bhopal workers.

THE ACCIDENT

The conference organisers invited Dr. Ved Prakash Gupta, Director of the Indian Central Labour Institute, to deliver a paper on the Bhopal disaster.

Unfortunately, circumstances have prevented Dr. Gupta from participating in our conference. The Government of India has not completed the investigation, there are a number of substantial compensation claims against the company and Government officers are not free to talk about some aspects of the disaster in public, at least at present. Because of the numerous personal discussions, involving preconditions for the disaster, with a number of Indian labour inspectors and other Government officers who participated in the Bhopal disaster investigation, I accepted the invitation to prepare this paper. However, at least in its written section, the paper cannot be substantially in variance with the reports which have been already published (see References 1 and 2 on page 25), on which this paper is based, without giving away the sources of confidential information.

MIC is highly reactive, unstable, flammable, volatile and toxic. It reacts with acids, alkalies, water and a variety of organic chemicals. Most of these reactions are exothermic (they give off heat), and some are violent. The flashpoint of MIC is 18°C, and a concentration of only 6 per cent in air is explosive. MIC boils at 39.1°C. The threshold limit value set by the American Conference of Governmental Industrial Hygienists is 0.02 ppm — among the lowest for any substance. Union Carbide's material safety data sheet states: "Methyl isocyanate can undergo a 'run-away' reaction if contaminated. A vapour cloud constitutes hazards from the standpoint of ignition (a 'fire-ball' could result) and toxicity".

Such a chemical deserves the highest standard of safety involved, and the Bhopal MIC plant contained several safety systems. MIC reacts more violently when warm, so the storage tanks included refrigeration systems. The tanks were protected from overpressure by safety valves and rupture discs. The system was designed to vent escaping gases to a vent gas scrubber, to be neutralised with caustic soda, and then to a flare tower. But even such a well-designed system could not do the job without being used properly and without being supplemented by well-designed and meticulously implemented safe working procedures.

The Union Carbide report states that the reaction in the storage tank which released the lethal gas was triggered by 1,000-2,000 lbs. of water entering the tank.

Based on the information gathered by the investigators, several hypotheses and scenarios were proposed and examined for compatibility with known facts. They

20

were tested experimentally, where necessary, for compatibility with pertinent analytical and chemical data. A single scenario to fit the event could not be proposed with complete certainty, since sufficient critical information was missing in each case due to restrictions placed on the investigators.

One-to-two thousand pounds of water entered the tank. Although entry from vent headers cannot be ruled out at this time, direct introduction of water through the vent or other piping has a higher probability for occurrence closed valves. The large amount of water was necessary to generate the heat needed to initiate and accelerate the subsequent reactions. The temperature of MIC in the tank before the incident was at 15 to 20°C, as compared to the requirement of about 0°C. The lower temperature would have retarded the reaction rates and considerably extended the time available for corrective action. The refrigeration system provided to cool the MIC in the storage tanks had been made non-operational in June 1984.

The reaction of water with MIC led to an increase in pressure due to evolution of carbon dioxide as well as an increase in temperature due to the exothermicity. The increase in pressure to 10 psig, noticed by an operator at 11.00 p.m. on December 2, is believed to be due to this phenomenon. The increase in temperature was not signalled by the tank high-temperature alarm, since it had not been reset to a temperature above the storage temperature. At 12.15 a.m. the operator checked the tank pressure again. It was 30 psig., and rising rapidly. Seconds later, the reading was off-scale. The rupture disc/safety valve system is designed to give way at 40 psig, and when it did, the contents of the tank rushed through the lines at a rate of almost 10,000 pounds per hour, later reaching a rate of about 40,000 pounds per hour, owing to increased pressure and temperature — 200°C and 200 psig — resulting from continuing exothermicity.

The escaping gas went first to the vent gas scrubber. It is not yet clear whether the scrubber functioned on the night of the release. The pump had been shut off and the instrument panel in the control room indicated that it could not be restarted. On the other hand, the caustic soda in the scrubber was found to be hot the next morning, indicating that some reaction had taken place. In any event, from the information published in the Bhopal operating manual and press reports after the accident, it appears that the scrubber did not have the capacity to handle the massive release completely. From the scrubber, the gas should have gone to the flare tower, but the unit was out of service. The pipe leading to it had been removed for maintenance weeks earlier. So the gas was vented directly to the atmosphere (Appendix I gives the Layout of MIC Tank, Vent Scrubber and Flare Tower). Several workers made a last-ditch effort to spray the escaping gas with water to neutralise it. Wearing full-face respirators and rapidly running out of air, they struggled with poorly maintained valves to turn on the water spray, only to find that the pressure was not sufficient to reach the gas. By 1.00 a.m. on December 3, a lethal cloud was drifting over the unsuspecting neighbourhoods of Bhopal, where it would kill at least 2,500 people and injure more than 200,000.

The Bhopal plant had four major safety systems designed to prevent or neutralise an uncontrolled MIC reaction:

(a)   a 30-ton refrigeration unit to cool stored MIC, in order to prevent it from vapourising or reacting;

(b)   a vent gas scrubber to neutralise toxic gases with caustic soda in the event of a release;

(c)   a flare tower to burn vented gases from the MIC storage tanks and other equipment; and

(d)   a water spray system to knock down escaping vapours.

At the time of the accident, at least three of these systems were not operating:

(a)   The 30-ton refrigeration unit had been shut down since June 1984. There were no mechanical problems with the system — it was taken out of service to save money. The Freon refrigerant had been drained out for use elsewhere in the plant. The shut-down was in violation of established operating procedures.

(b)   The vent gas scrubber was turned off in October 1984, apparently because the supervisors thought it was not necessary when MIC was only being stored and not produced. In addition, the caustic flow indicator was malfunctioning, so it would have been difficult to verify whether the unit was operating or not.

(c)   The flare tower had also been out of service since mid-October. A section of corroded pipe leading to it had been removed, even though the replacement pipe was not ready. In addition, another procedure failure had compromised the reliability of the flare tower even before it was disconnected. The tower was originally built with a back-up set of fuel gas cylinders to ensure that the pilot light stayed on. However, the back-up system was discontinued to save money.

According to the workers interviewed, the water spray shroud, which was activated the night of the accident, did not reach the level of the gas release and was, therefore, useless. In 1982, Union Carbide Corporation, after inspecting the Bhopal facility, had recommended a new, larger water spray system, but it was never installed.

Inadequate maintenance was a long-standing complaint at the Bhopal plant. The poor maintenance of the major safety systems has already been mentioned. These problems extended to production equipment.

Leaking valves and malfunctioning gauges were common throughout the facility. A 1982 Union Carbide Corporation inspection of the plant by U.S. safety personnel noted such problems, and resulted in the replacement of valves in the MIC unit, but at the time of the accident, valves and pipes had again corroded and leaking valves were a serious problem. Leaking valves probably allowed water to enter the tank. Broken gauges made it hard for MIC operators to understand what was happening. In particular, the pressure indicator/control, temperature indicator and level indicator for the MIC storage tanks had been malfunctioning for more than a year.

At the time of the accident, the Bhopal plant, including the MIC facility, was operating with reduced manpower. According to the workers and published reports, the plant had been losing money, and in 1983 and 1984 there were more personnel reductions in order to cut costs. Some workers were urged to take early retirement, 300 temporary workers were laid-off and 150 permanent workers were pooled and assigned to jobs, as needed. The workers said that employees were often assigned to jobs they were not qualified to do. This practice was also noted by Union Carbide Corporation in its 1982 inspection report.

According to the workers, the maintenance supervisor position on the second and third shifts had been cut on November 26, less than a week before the accident. The maintenance supervisor was responsible for preparing all maintenance work and for giving instructions for the job. The workers indicated that it would have been the responsibility of the maintenance supervisor to prepare the pipe which was being flushed with water on the night of 2 December 1984. This is suspected to be one possible source of the entry of water into the pipes leading to the MIC storage tank, perhaps responsible for the runaway chemical reaction.

When the post of maintenance supervisor was eliminated, these responsibilities apparently shifted to the production supervisor. But the production supervisor on duty on the night of 2 December had been transferred from the Union Carbide Battery Plant one month before and was not fully familiar with either the operating or maintenance procedures.

Training was a major problem at the Bhopal plant. At the time the MIC facility was opened in 1980, 25 people were sent to the United States for training. But due to high turnover — around 80 per cent in the MIC plant from 1982-1984 — few of the people originally trained in MIC operations in the U.S. remained in Bhopal.

The workers said that they had been given little or no training about the safety and health hazards of MIC or other toxic substances in the plant: they thought the worst effect of MIC was irritation of the eyes.

Union Carbide was aware that the chemicals produced and used in the Bhopal plant posed a risk to workers and the community at large. Leaks of toxic chemicals from the plant had effected both the workers and the public. However, a disaster plan for warning and evacuating the community in the event of a leak had never been developed. The plant is reported to have had two sirens. One to warn workers in the plant and one to warn the public were sounded — some press accounts report a delay of several hours. Even when the alarms were sounded, people affected by the gas had no idea of what was happening or where they should flee.

Emergency plans for workers in the plant were only marginally better. In the event of a leak, the workers had been instructed to check the wind indicators and run into the wind, away from the direction of the gas dispersion. However, most of the escape routes from the plant were blocked. The plant is surrounded by an 8 foot concrete wall, topped with barbed wire. Only one gate was kept open. Since the gas was present in the area of the gate, workers were forced to climb over the wall and squeeze through the barbed wire to escape.

In addition to the underlying reasons (preconditions) and the immediate direct causes described above, failure to inform workers and the public contributed to the accident and increased its severity. The Bhopal plant produced, used, stored and transported a number of toxic and hazardous pesticides, feedstock and chemical intermediates that posed a risk to the workers and the public. However, the company had never provided complete information about these chemicals to the workers, Government authorities or community residents.

Even after the accident, the company failed to provide adequate information on MIC and its hazards. On 3 December, as thousands of people lay dead or dying on the streets, the Medical Director of the Bhopal plant continued to insist that methyl isocyanate was only an irritant and not life-threatening.

Medical authorities stated that they received little, if any, information from the company on the diagnosis and treatment of MIC injuries. A hospital Director said that he finally found out that the chemical was MIC from a newspaper report on the evening of 3 December. The State Health Director finally received solid information on the chemical from the World Health Organisation several days after the accident.

CONCLUSIONS

The Bhopal disaster was caused by a combination of factors, including the long-term storage of MIC in the plant; the potentially undersized vent gas scrubber; shut-down of the MIC refrigeration units; the company's failure to repair the flare tower, leaking valves and broken gauges; and cuts in manning levels, crew size, worker training and skilled supervision. The accident might have been prevented if the plant had done more to follow up its 1982 safety inspection. The effects of the accident were magnified by the company's failure to provide adequate information to its subsidiary, authorities and community residents; siting of the plant close to residential areas; and lack of disaster planning.

The combination of contributory factors which caused the disaster on the night of 2 December 1984 was unique. But the underlying reasons (preconditions) for the Bhopal disaster, i.e.

(a)   insufficient attention to safety in the process/operations design;

(b)   unsafe operating procedures;

(c)   inadequate maintenance, leading to faulty equipment;

(d)   cutbacks in manning and inadequate training;

(e)   lack of skilled supervision;

(f)   siting of potentially dangerous plants in heavily-populated areas;

(g)   lack of information; and

(h)   lack of planning

are in existence in a number of plants and factories in different countries around the world.

All the above factors are most clearly examples of failure in the sphere of organisational and individual behaviour, or human error. It is the aim of this conference to highlight such human errors, in order to minimise human suffering and economic losses resulting from occupational accidents and discuss the ways they can be avoided.

REFERENCES:

1.   Bhopal Methyl Isocyanate Incident Investigation Team Report, Union Carbide Corporation, Connecticut, March 1985.

2.   The Trade Union Report on Bhopal, ICFTU/ICEF, Geneva, July 1985.

3.   Jain, N.K., Assistant Director, Industrial Health and Safety, Bhopal. Personal Communication.

4.   Murashetty, M., Director, Safety, Central Labour Institute, Bombay. Personal Communication.

LAYOUT OF MIC TANK, VENT SCRUBBER AND FLARE TOWER

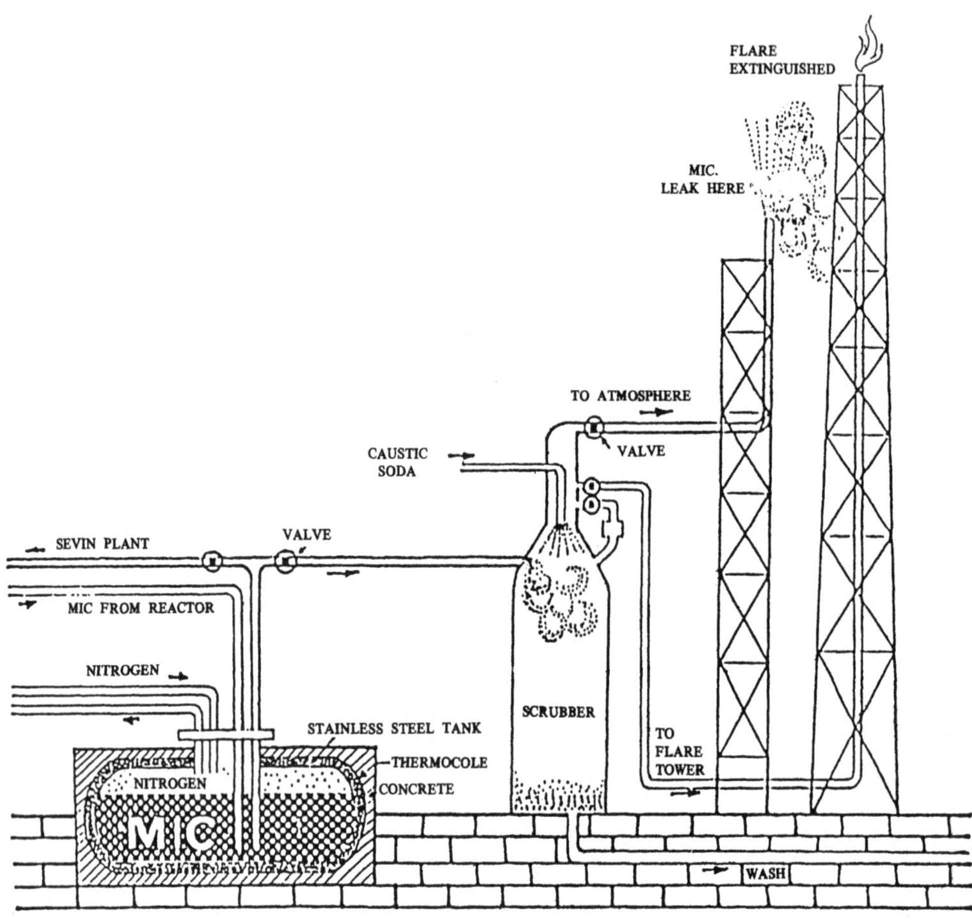

PHYSICAL HAZARDS: FIRES AND EXPLOSIONS
by
Milos Nedved

All of us have already seen a fire. Most of us have experienced some kind of explosion, at least a very minor one. Sometimes fire is a very useful phenomena, as e.g. in an oven, a furnace, a boiler, etc. On the other hand, unwanted fires cost us hundreds of lives every year, and damages caused by fires and explosions are in the order of hundreds of millions of pounds.

Effective fire prevention and protection, based on an understanding of combustion phenomena, is a very important, integral part of environmental health and safety. This chapter introduces readers to the subject at a level enabling them to evaluate and minimise hazards in the workplace, and to understand more advanced texts on fire prevention and protection, some of which are listed at the end of the chapter (under References).

WHAT IS A FIRE?

A fire is usually unwanted, and often uncontrolled, combustion of a fuel, in most cases in air, and usually is accompanied by flame. We all have quite a clear idea of what is meant by a flame, but it is difficult to give the word a precise and a short definition. The Oxford Dictionary gives: "portion of ignited gas; visible combustion".

Combustion processes are exceedingly exothermic chemical reactions, i.e. the reactions give off heat and, because of the high temperatures involved, they are mostly vapour phase reactions, that is reactions between two or more gases, one of which is usually oxygen. There are exceptions to this rule, like the smouldering or burning of metals, which will be discussed later.

IS THERE ANY SIMPLE BASIC MODEL OF A COMBUSTION PROCESS?

If a molecule of fuel comes into very close contact with a molecule of oxidant (usually oxygen), a chemical reaction takes place, but not always. Collision of the two above molecules connected with low energies does not mean a chemical reaction — the molecules will rebound elastically and nothing happens. If, however, the two molecules collide with sufficient energy, a chemical reaction takes place and combustion products are formed. As this is an exothermic process, a certain amount of heat energy is formed. This energy can then heat up the next set of fuel/oxidant molecules.

The energy of individual molecules depends normally on temperature — the higher the temperature the higher the energy of the molecules. From these bits of information we can already see that a combustion process requires three components: a fuel, an oxidant (usually oxygen in air) and high temperature (an ignition source). If all these three components occur simultaneously, a fire breaks out. If one of the components is removed, a fire is extinguished. This piece of knowledge — so-called 'fire triangle' — enables us to understand fire prevention (how to prevent fires from starting) and fire protection (how to fight fires whose occurrence we have not been able to prevent).

WHAT ARE THE MOST IMPORTANT POINTS IN FIRE PREVENTION?

We already understand that a combustion process can be maintained in a flame only when sufficient heat energy is being transferred to the unburnt gases to increase their temperature high enough for a chemical reaction to occur — so called flame temperature. This fresh gas entering the flame can then react, and the process goes on. The flame is most stable when there is just enough oxygen to burn up all the fuel. This is the so-called stoichiometric composition of fuel/oxygen mixture and can be easily calculated from the chemical equation describing the combustion process. For example, the stoichiometric composition of pentane/air mixture is 2.5 per cent of pentane in air. Burning this mixture, we can get the hottest pentane/air flame, and the flame temperature will be progressively lower if we burn lower or higher concentrations of pentane in air. The reason is that some gases (either air or fuel — always those which are in excess of a stoichiometric composition) are being heated up to flame temperature and then do not contribute to the reaction — this portion of heat energy is effectively wasted. Because of this waste of energy, pentane will not burn at all if the mixture contains less than 1.5 per cent of pentane or more than 7.8 per cent. We describe this situation by stating that the lower limit of flammability (LL) of pentane in air is 1.5 per cent and the upper limit of flammability (UL) is 7.8 per cent. Limits of flammability can be easily measured in the apparatus designed for this purpose and the vast majority of industrially-used flammables are tabulated [Sax (1979), Fire Protection Association (1974)]. Some examples are listed in the table on page 29.

HOW CAN WE MAKE USE OF THESE AND SIMILAR DATA?

To use the above data for fire prevention, we can see quite clearly that fire, or explosion of e.g. acetone vapour in air, could occur only if the atmosphere in a workshop contains between 2.6 volume per cent and 12.8 volume per cent of acetone. Our task would be, therefore, to maintain the content of acetone vapour well below the lower limit of flammability, by means of efficient ventilation. At the same time, to be on the safe side, we can apply our knowledge of the fire triangle by eliminating all the possible ignition sources.

Limits of Flammability of Some Gases and Vapours in Air

Fuel	*LL [%]*	*UL [%]*
Methane	5.0	15.0
Ethane	3.0	12.5
Propane	2.2	9.5
Butane	1.9	8.5
Pentane	1.5	7.8
Acetone	2.6	12.8
Petrol (typical composition)	1.3	6.0
Ethylalcohol	3.3	19.0
Acetylene	2.5	100
Hydrogen	4.0	75

DO ALL FLAMMABLE LIQUIDS EASILY FORM EXPLOSIVE MIXTURES?

If we recall the basic model of a combustion process, we remember that a certain concentration of fuel is required for a fire — at least the lower limit of flammability. But not all flammable liquids are capable of giving off enough vapour (at a certain temperature) to reach this lower limit. Some liquids are more volatile, while some are less volatile, e.g. hexane, at room temperature, releases so much vapour that flammable mixtures can be formed at some distance from the liquid. Paraffin, on the other hand, is much less volatile, and at room temperature it is impossible to reach the concentration of lower limit of flammability to form even very near to a liquid surface. At the same time, we know that the higher the ambient temperature, the more vapour is given off by the liquid.

By heating up liquid paraffin, more and more vapour is generated above the liquid surface. At 55°C its concentration is just equal to the lower limit of flammability, and if an ignition source is applied, we have all the three necessary components for a fire to start — a flame will now flash through this vapour/air mixture. This temperature (55°C for paraffin) is, therefore, called flashpoint. If you like, you can already write down a scientific definition: flashpoint is the lowest temperature at which a liquid will give off enough flammable vapour at or near its surface, such that its mixture with air can be ignited by a spark or flame.

IS FLASHPOINT AN IMPORTANT PIECE OF DATA?

This is a very useful piece of information, being sometimes of utmost importance for fire prevention. A liquid with a flashpoint below the normal ambient temperature presents a much greater hazard than one with a much higher value. If we are handling or storing flammable materials, we must always do this in such a way as to prevent the formation of a flammable mixture — and this depends on the value of the flashpoint of a liquid. Some liquids have very low flashpoints (petrol $-43°C$, acetone $-18°C$, ethylalcohol $+13°C$, hexane $-22°C$). It means that at normal temperatures enough vapour is given off to form a flammable mixture, and some other methods (effective ventilation and elimination of all ignition sources) must be used to control the hazards. On the other hand, the storage of paraffin, with its flashpoint equal to $55°C$, is quite safe at normal temperatures, and if we keep the temperature in the storage area well below $55°C$, flammable mixture cannot be formed. Another example of a flammable liquid with a relatively high flashpoint is ethylene glycol (f.p. $111°C$), which is used as a main component of some anti-freeze mixtures. Methylalcohol, which is used in some other anti-freeze mixtures, has f.p. $+10°C$. You can now see the difference in degrees of hazard connected with the use of the above mixtures yourself.

At a little higher temperature than flashpoint, the flame will persist long enough for continuous burning to be established (compared with a flashpoint at which a flame will only flash through the flammable mixture). This higher temperature, called fire point, is again a very important characteristic for each flammable material and can be defined as the lowest temperature at which a mixture of vapour and air continue to burn when ignited. Differences between the flashpoint and fire point of individual flammable liquids are usually $20\text{-}30°C$. In older English literature, fire point is rarely quoted, but American and newer English literature usually quote both flashpoint and fire point. All these values are easily accessible [(e.g. Sax (1979), Fire Protection Association (1974)].

The next step will be to have a look at another component required for starting a fire — ignition sources.

WHY IS UNDERSTANDING OF IGNITION SO IMPORTANT?

In certain situations we are not able to eliminate the presence of fuel — it can be the prime aim of our activities, as e.g. in the chemical or petrochemical industries. Also, very frequently we cannot eliminate the presence of air, because without air people would suffocate. What we can, however, do is to eliminate or minimise ignition sources in the area, for which fire prevention is being considered.

IS THE ELIMINATION OF IGNITION SOURCES AN EASY JOB?

Unfortunately, an examination of case histories shows that there are many ways of igniting a flammable mixture. The commonly quoted ignition sources [e.g. Bradley (1969), Lewis and Elbe (1961), and Wharry and Hirst (1974)] include:

(a)   naked flames;

(b)   electrical sparks from power sources;

30

(c)   electrical sparks from electrostatic discharges;

(d)   hot surfaces;

(e)   frictional sparks;

(f)   spontaneous ignition;

(g)   pyrophoric materials;

(h)   compression of flammable mixtures; and

(i)   radiation.

This list cannot be regarded as fully comprehensive, and experience suggests that the complete elimination of all possible ignition sources is very difficult, and sometimes impossible. To minimise possible ignition sources effectively in the varied circumstances that arise in real life, it is necessary to understand the main factors that control ignition, and on which fire prevention is based.

WHAT ARE THE MOST COMMONLY OCCURRING IGNITION SOURCES?

An examination of the above list of ignition sources shows only three basically different ignition sources:

(a)   naked flame;

(b)   spark ignition; and

(c)   hot surfaces.

All the remaining sources in our first list can, because of similarity of ignition mechanisms, be grouped under the above three headings.

There is no need to discuss ignition by naked flame in detail, since virtually any flame will ignite a flammable mixture. To prevent ignition, the two must be kept well apart.

The remaining two categories show much more variety in their capability for ignition. They will now be discussed separately in more detail.

CAN A SPARK IGNITE A FLAMMABLE MIXTURE?

Considering again our basic model of a combustion process, we can see that a certain minimum amount of energy must be applied to a certain minimum volume of fuel/air mixture before a fire can break out. This energy, so called minimum ignition energy, differs from one flammable mixture to another — for mixtures of most hydro-carbon vapours with air, the minimum ignition energy is about 0.2 mJ (mJ = $\frac{1}{1,000}$ joules). Values for other organic fuels do not differ very much from this, but hydrogen/air mixture can be already ignited by a spark of 0.02 mJ. Values are much smaller for mixtures with oxygen — the minimum ignition energy for most hydro-carbon/oxygen mixtures is 0.002 mJ.

From these data we can see that many, but not all sparks can ignite flammable mixtures. Knowledge of the conditions where there is no ignition allows certain correctly designed electrical apparatuses to be used in hazardous conditions — we then say that an item of electrical equipment is intrinsically safe.

Intrinsically safe equipment only incorporate items which have low power requirements. For other electrical equipment, a variety of techniques are available — for example, segregation, intended to keep the flammable atmosphere away from the potential ignition source, various forms of enclosure and non-sparking devices.

Electrical sparks sufficient to be an ignition source can also result from an electrostatic discharge.

HOW IS AN ELECTROSTATIC SPARK CREATED?

Whenever a substance (solid, liquid or gas) is sliding over a different substance, electrons can be rubbed off one of them and collected on the other. Usually they can find their way back again quite easily, but sometimes if one or both substances are good electrical insulators, large numbers of electrons can become separated from their original atoms. The surface with the surplus of electrons has a negative charge, and the one that has lost them has a positive charge. When this electrical charge builds up to a certain voltage, causing a breakdown of the insulating properties of the medium (usually air it discharges itself spontaneously), a spark will pass as all the lost electrons return back to their original positions.

Hazardous electrostatic discharges are readily produced by the pumping of organic liquids, flow of powders, discharge of steam and flow of air in an extract ventilation system.

CAN WE ELIMINATE THE FORMATION OF ELECTROSTATIC SPARKS?

In situations where electrostatic discharges can be very hazardous, special precautions must be taken. Since it is usually very difficult to prevent the separation of electrical charges, the safety measures adopted are directed towards either the reunion of charges or the elimination of all flammable mixtures from areas where electrostatic discharges could occur. The simplest method of the speedy reunion of the charges is by earthing all conducting parts of the plant.

In an oil refinery, flammable liquids of very low conductivity (e.g. petrol) are pumped at low velocities to minimise the charge that is generated. In explosives factories, people handling sensitive explosives may be earthed either through conductive footwear or by having their hands in contact with an earthed metal plate. In hospitals, where mixtures of flammable anaesthetics and oxygen are given to patients, operating theatres may have conducting floors and staff wear conductive footwear. The most up-to-date technique of preventing the electrostatic buildup is to use a source of ionising radiation. The atmosphere then becomes conducting and the charges are reunited without an electrical spark. However, the use of ionising radiation is connected with other complications (protection of people), therefore this technique is used only where other classical methods cannot be used.

Frictional sparks, such as are produced when a piece of metal or a stone is struck a blow with a hammer, are also capable of igniting flammable gas mixtures. Again, a minimum volume of fuel/air mixture must be heated up and sufficient energy supplied by the hot fragments. This leads us to the second very important category of ignition sources — hot surfaces.

WHAT TYPES OF HOT SURFACES CAN ACT AS IGNITION SOURCES?

In practice, hot surfaces vary widely in size, shape and surface material, as well as in temperature. These factors, in addition to the properties of the flammable mixture, will all influence whether ignition occurs or not. Actual surfaces vary from large areas, as on furnaces or heated lines, through vehicle engines and overheated pumps, to small sources, such as electrical fire filaments and sparks from cutting and welding operations.

In general, hot surfaces have to be at a significantly higher temperature than the auto-ignition temperature (AIT) of the flammable material before ignition can occur. Auto-ignition temperature is the temperature at which a material will self-ignite and sustain combustion in the absence of an outside flame. Many lists of AIT are available [e.g. Hilado and Sharp (1972)]. Some organic chemicals have very low values of AIT, e.g. ethyl nitrite, 90°C; carbon disulphide, 102°C; and acetaldehyde, 140°C. However, the temperature needed to ignite is lower if the contact time is extended, or if the gases are stirred. Because of the possible catalytic reactions between the flammable mixture and the hot surface, ignition may even occur at a temperature lower than the auto-ignition temperature. If, in a car accident, petrol is spilled and the petrol/air mixture comes into contact with the hot exhaust, it may ignite after a delay of several minutes, although the temperature of the hot exhaust is only about 200°C.

Hot temperatures can also frequently result from spontaneous heating.

WHAT ARE THE CAUSES OF SPONTANEOUS HEATING?

The hardening of paint, which takes place after all the volatile solvents have evaporated, is due to oxidation by air oxygen. There are oils in the paint that are accepting oxygen atoms and then polymerising. A certain amount of heat is released during this process, but this is quickly lost to the surrounding atmosphere, so the temperature increase is negligible. However, if we use some rags to wipe up the spilled paint or to clean the brushes, and then throw the rags into a heap of rubbish, the situation is quite different. A rag is a good thermal insulator. The thin layer of paint on the rag reacts with oxygen and heat is released, which cannot easily escape. The temperature begins to rise, and may continue to rise until the material in the middle of the heap begins to smoulder. This process is called spontaneous heating, which will raise the temperature of a solid material to its auto-ignition temperature, and flaming combustion takes place. The whole process is then called spontaneous ignition.

Hay provides another example of the same phenomenon. The temperature is raised by the action of micro-organisms, and in the well-insulated interior of a haystack, the temperature continues to rise up to spontaneous ignition. Some spontaneous ignition temperatures are listed below:

Spontaneous Ignition Temperatures

Material	*Spontaneous Ignition Temperature* $^{\circ}C$
Coal	125
Hay	170
Newspaper	185
Sawdust	195
Cotton	225

WHAT MATERIALS ARE CAPABLE OF SPONTANEOUS HEATING, AND WHEN?

Almost all combustible materials are capable of spontaneous heating when their temperatures have been increased near to the auto-ignition temperature. Examples of this phenomena, well-known in industry, are lagging fires. Some pipelines in chemical plants are lagged, or insulated, to prevent heat losses. If a spilled combustible liquid. like fuel oil or others, penetrates into the lagging, it is quite possible that after a delay of perhaps several hours, or days, the contaminated lagging will be smouldering. This represents a serious hazard in areas where special care has been taken to eliminate all ignition sources. From case histories we can see that this is one of the commonest causes of fire in the chemical and petrochemical industries.

Other hazardous materials subject to spontaneous heating are coal, charcoal, fish meals, fish oils, linseed oil, soya bean oil, soap powders, etc. All of these can ignite spontaneously, if the right precautions are not taken.

The main emphasis, in this rather lengthy section of the chapter, has been to show the extremely wide variety of ignition sources we can encounter. Experience teaches us that a fire triangle can be modified: fuel and air can frequently mean fire, since ignition sources are always with us, even if we are not always aware of their existence.

34

IS THERE A DIFFERENCE IN THE BURNING OF SOLIDS, LIQUIDS AND GASES?

So far, we have been talking mainly about the ignition of mixed gases, and indeed this is the way how most combustion processes start. If we want to light a candle, the flame has to melt and vaporise some wax in the wick. This vapour makes a mixture with air, which is within the limits of flammability. The mixture is then ignited by the wick, and a flame is established around. Some solids, like wax or polyethylene, will first melt and then vaporise. Others, like paper and wood, produce vapours directly by pyrolysis. The large complex molecules in wood are pyrolysed to form simpler gaseous molecules, like methyl alcohol, hydrogen and carbon monoxide.

Since solid materials need first to be pyrolysed, they do not burn so readily and as fast as flammable liquids or gases. Knowledge of the burning characteristic of solids, liquids and gases will enable us to fight fires more effectively.

Some solids can burn in the absence of flame, by smouldering. The rate of propagation is usually very slow (e.g. for wood sawdust, the smouldering rate is in the region of 10-20 cm./hour). The burning zone is frequently covered by a layer of ash and, therefore, it is not easily visible. If smouldering comes into contact with more flammable materials, such as oily rags, paper, wood shavings, etc., flame combustion will follow. There may be a long delay between the initiation of smouldering by an ignition source, such as sparks from welding or cutting operations, and the outburst of flaming. This delay may be of the order of days. Now we can see quite clearly why careful supervision of cutting, welding and grinding operations is necessary to prevent undetected sparks from falling in between the floor-boards, where there is always a certain amount of combustible dust.

The burning of solids with flame depends on the size and shape of the material. The thinner the material and the lower its density, the faster it will burn. A finely divided combustible solid can burn very rapidly, to give a dust explosion. This case will be discussed later.

Liquids burn much faster than solids, and the rate of flame-spread over a liquid surface depends on the flashpoints.

The lower the flashpoint, the higher the horizontal rate of the spread of flame. The rate of burning vertically downwards is, however, relatively independent of the flashpoint, and is fairly constant for a wide range of flammable liquids.

The burning velocities of gases in flammable mixtures are very high — we are frequently talking about explosions.

WHAT IS MEANT BY FIRE PROTECTION?

Two main steps to take are the prevention of ignition and the rapid control and extinguishing of a fire. As we have already seen before, fires involving different materials will behave in different ways. Therefore, it is necessary to use different ways of fire-fighting and different extinguishing agents.

For these purposes, fires are classified as follows:

Classification of Fires

Class	Material Involved	
	Continental Classification	*U.S. Classification*
A	Solids (excluding metals)	Solids
B	Liquids	Liquids
C	Gases	Energised electrical equipment
D	Metals	Metals

As we can see, there are two classifications in use, differing only very slightly. Each of the classes of fire, owing to the different chemical nature of the combustible involved, requires a different extinguishing agent.

WHAT ARE THE MOST IMPORTANT EXTINGUISHING AGENTS?

The extinguishing agents used most commonly nowadays are water, foams, inert gas, vaporising liquids and dry powders.

WHAT ARE THE ADVANTAGES AND DISADVANTAGES OF THESE EXTINGUISHING AGENTS?

Water is cheap, readily available, non-toxic and has a high specific heat and heat of evaporation. Its disadvantages are that it conducts electricity, cannot be used against reactive metals and is not miscible with some organic liquids. It is mainly used in extinguishing Class A fires. Water is usually applied either as a jet or as a spray.

When 1 volume of water evaporates at $100°C$, approximately 1,600 volumes of water vapour are formed. The modes of action in extinguishing fires are:

(a)   blanketing fires by steam, to exclude air;

(b)   cooling of flames and surfaces; and

(c)   dilution of mixible liquids.

Foams are used mainly for fighting fires in flammable liquids, or to fill enclosures where access in difficult. Foam concentrate in water is aerated and, for the high expansion foams, 1 volume of the concentrate in water produces up to 1,000 volumes of this modern extinguishing agent. The modern foams, based on fluorinated chemicals, have a film-forming property. It enables not only the cooling down of the burning material, but also the impermeable film, formed on the surface, prevents re-ignition after the evaporation of water. If used against flammable liquids, the film on the liquid surface reduces evaporation and, therefore, hinders burning. It also cuts off further oxygen supply. High expansion foams can be advantageously used to completely fill e.g. cellars or basement stores, to avoid having to send in fire-fighters.

Inert gas is usually carbon dioxide. In enclosures, it is more effective than against the open fires.

Vaporising liquids (halogenated hydrocarbons) are capable of active chemical interference with combustion processes, and are very effective from the point of the weight needed to extinguish the fire. However, they tend to be more expensive. They are the best agents for dealing with fires in electrical equipment because they are non-conductors and penetrate the spaces inside equipment.

Dry powders (or dry chemicals) are effective agents against liquid spillage fires. They can extinguish very large fires, but can obstruct the vision of fire-fighters. Their mode of action involves both chemical and physical mechanisms. After the powder cloud has extinguished the flames on a burning solid, it melts and forms a film on the material. The melting has some cooling effect and the film has a smothering effect, reducing the rate at which flammable vapours are produced from the material.

Foams, dry powders, carbon dioxide and halogenated hydrocarbons are all used in portable extinguishers. The suitability of various extinguishing agents for fires of different classes is summarised below:

Choice of Extinguishing Agent

	Water	*Dry Powder*	*Carbon Dioxide*	*Foam*
Class A Wood, cloth, paper and similar combustible materials	Most suitable	Not recommended, except for small surface fires	Not recommended, except for small surface fires	Not recommended, except for small surface fires
Class B Flammable liquids – petrol, oils, greases, fats	Unsuitable	Most suitable for general use	Most suitable where contamination by deposits must be avoided	Most suitable where re-ignition risk is high
Class C Electrical plants	Unsuitable – dangerous	Suitable	Suitable	Unsuitable – dangerous

To enable the operator to see clearly, without lengthy reading, what the contents of a portable extinguisher are, certain colours on extinguishers indicate the extinguishing agent. These are as follows:

Colours on Extinguishers Indicating Extinguishing Agent

Extinguishing Agent	*Colour*
Water	red
Foam	beige
Carbon dioxide	black
Dry powders	blue
Vaporising liquids	dark green

The extinguishing agents and fire-fighting methods are extensively described in numerous literature (for example, the Manual of Firemanship, Books 1-18).

WHAT ARE THE OTHER DANGEROUS EFFECTS OF FIRES?

In addition to the heat generated by fires, formation of smoke and toxic gases are additional hazards. Excessive smoke can hinder the escape of persons and hamper the activities of fire-fighters.

Toxic gases generated by fires are particularly dangerous. Carbon monoxide, which is most frequently produced by carbonaceous materials, is odourless and can overcome people during the first stage of a fire. Statistics show that much more people die in fires as a result of being overcome by toxic products than by the direct action of heat or falling walls and ceilings. Elements, such as chlorine, sulphur and nitrogen, present in a combustible material are causes of formation of irritant gases. PVC involved in a fire releases up to half of its weight as an irritant hydrogen chloride gas. Polymers containing nitrogen (e.g. polyurethane foam) can generate hazardous concentrations of extremely toxic hydrogen cyanide. Here, we can clearly see the close relationship between fire prevention, fire protection and human safety (toxicology) within the framework of the multidisciplinary subject of safety and hygiene. To illustrate and emphasise these points, data on formation of toxic products (carbon monoxide and hydrogen cyanide) from some very commonly used flammable materials is given below:

Toxic Products of Burning Materials

Material	Carbon monoxide Released mg/g of Sample	Hydrogen cyanide Released mg/g of Sample
Wool	0.230	—
Acrylic fibre	0.300	0.05
Cotton	0.500	—
Polyurethane foam	0.510	0.03

Since, nowadays, almost everybody wears some garments containing acrylic fibres, and polyurethane foam is used frequently in furniture, more and more attention should be paid to fire prevention to protect the health and property of both ourselves and our fellow citizens.

WHAT IS AN EXPLOSION?

The Chambers Dictionary defined an explosion as "a rapid increase of pressure in a confined space, generally caused by the occurrence of exothermic chemical reactions in which gases are produced in relatively large amounts". We can also use a simpler but wider definition: "a rapid release of energy". The sources of the energy may vary, from the heating of a tin of meat to the energy of a compressed gas, a nuclear device, a reaction of decomposition or thermal expansion of fast polymerisation. All these can give rise to explosions, but the energy release must be sudden. This energy is then dissipated, being used in various forms of external work, as fragment effect, shock wave generation, etc. According to the source of energy, explosions can be classified as follows:

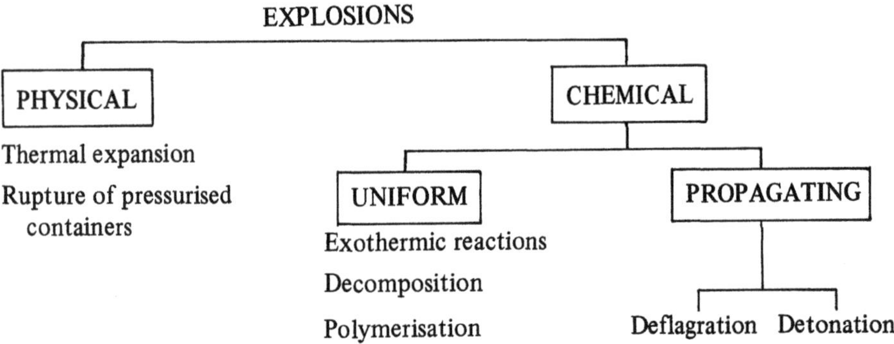

The propagating chemical explosion, which is one of major interest, arises when flammable gas/air or gas/oxygen mixtures within the limits of flammability are ignited. In case of detonation, the shock wave travels at supersonic velocity — a speed greater than that of sound — in the gas mixture. For deflagration, this velocity is significantly lower. Pressures in a detonation wave are much higher than pressures arising from deflagrations. A deflagration may produce a maximum explosion pressure of 7 atm, if contained, whereas a detonation may produce a peak pressure of about 20 atm.

For some fuel/air mixtures, both modes of combustion — deflagration and detonation — are possible. Which one occurs depends on several factors, e.g. strength of the initiating source, roughness, and length and diameter of the containing pipe or vessel, as well as the pressure, temperature and composition of the mixture. Examples of various fuel/air explosive mixtures have already been discussed.

HOW CAN WE SUCCESSFULLY AVOID EXPLOSIONS?

To prevent explosions from occurring, we must use basically the same techniques as described in fire prevention. The simultaneous occurrence of fuel, air and ignition sources would lead to an explosion, and must be avoided.

Particularly hazardous conditions can occur in an empty tank. The flammables remaining at the bottom of the tank evaporate, frequently forming a mixture of a composition within the limits of flammability. Violent explosions can then result from an ignition. Explosions of this kind are not confined to storage tanks. The spillage of a flammable liquid inside a building may also result in an explosion that wrecks the building. To prevent this from happening, great care must be taken to confine possible spills immediately, using e.g. sand or sawdust. These must then be disposed of safely, in an open space, far away from people and buildings.

Vapour/air explosions can also result when any solid fuel, like wood or plastic, has been smouldering in a closed room in insufficient air. Unburnt flammable vapours, resulting from pyrolysis of the solid fuel, may mix with air away from their source, and an explosion may propagate from an ignition source.

WHY ARE EXPLOSIONS INVOLVING DUSTS PARTICULARLY DANGEROUS?

Every year industrial dust explosions cause injury and death, and damage or destroy plants and buildings. Such explosions often start and spread fires.

WHAT IS A DUST EXPLOSION?

A dust explosion is the rapid combustion of particles of dust suspended in air. The heat produced by the combustion of these suspended particles causes rapid expansion of the surrounding gas, which in turn produces an increase in pressure at the walls of the vessel containing the dust cloud.

HOW CAN WE ASSESS IF A DUST EXPLOSION IN OUR PREMISES IS LIKELY?

We need to know:

(a)   if we process a dust that is explosible;

(b)   the minimum concentration of dust that will support an explosion;

(c)   the temperature at which the dust will ignite; and

(d)   the minimum ignition energy that will initiate an explosion.

WHAT ARE THE COMBUSTIBLE DUSTS?

Nearly all carbonaceous materials, if they are in sufficiently fine state, can give rise to dust explosions. Also certain metals can give rise to this phenomena, and generally metal dust explosions are much more violent than other dust explosions.

Generally speaking, the finer the dust, the greater its explosibility. It means the ease with which it is ignited is enhanced, and also the rate at which pressure rises in a confined explosion increases as the particle size decreases, and similarly the maximum pressure produced increases.

Listed below are some substances and processes subject to dust explosion:

Some Substances and Processes Subject to Dust Explosions

Substances	*Examples of Processes, Plant and Premises*
Aluminium	Blown powder production: settling chambers, cyclones, bag filter units, conveyors and buildings. Milled powder production: ball mills. Metal spraying: dust collectors and workrooms.
Cattle cake Cotton seed Feedstuff Sunflower seed cake Linseed	Milling machines, disintegrators, bucket elevators, cyclones and bag filter units; rotary drum dryers, storage bins and silos, and workrooms.
Coal	Milling machines, classifiers, cyclones, ducting and bag filter units in the production of pulverised fuel.
Cork	Milling machines, conveyors, screens; cyclones, bag filter units and workrooms.
Gramophone records	Milling machines and cyclones.
Magnesium	Hammer mills, ball mills, pan mills, conveyors, screens, cyclones and settling chambers (including improperly maintained scrubbers) associated with pulverising, abrading and scratching processes.
Malt	Mill and bucket elevators in distilleries; less serious explosions in breweries.
Phenol-formaldehyde resins	Mills, cyclones and settling chambers.
Polystyrene	Extruding machines, pneumatic dryer, cyclones, bag filter units, storage bins and workrooms.

Substances	Examples of Processes, Plant and Premises
Rubber	Dust chambers and collecting plant associated with abrasive machines used for tyre treading and retreading operations.
Starch	Starch powder production: disintegrators, cyclones, bag filter units, bucket elevators and pneumatic dryers. Starch moulding of sugar confectionery: workroom.
Wood flour	Milling plant, cyclones, screens and storage bins, associated with pulverising operations.

DO DUST CLOUDS BEHAVE IN A SIMILAR WAY AS MIXTURES OF FLAMMABLE VAPOURS WITH AIR?

Dust clouds do exhibit the phenomena of explosion limits. In most cases, the lower explosive limit is clearly defined, whereas the upper explosive limit is not so accurately defined.

The greater the moisture content of the dust, the lower its explosibility. However, the effect of moisture is not as great as people often assume, and we have had cases where starch dust containing as much as 17 per cent moisture has been involved in an explosion. Inert material present in dust will prevent an explosion, and there is experimental evidence of this: for example, the maximum pressure curve of various mixtures of starch and calcium carbonate shows a gradual descent, until a limit mixture of 60 per cent calcium carbonate and 40 per cent starch was unable to be exploded.

Reducing the amount of oxygen to prevent the ignition of a dust cloud is sometimes necessary in industry. Experimental tests to discover the limit at which propagation does not occur in various percentages of oxygen have shown that for the most carbonaceous dusts this figure is roughly half the normal oxygen content in the air. The use of inert gas for certain industrial processes does give very effective safety, but this kind of application is usually limited to firms and processes where a large supply of inert gas is readily available.

WHAT ARE THE MOST COMMON IGNITION SOURCES CAUSING DUST EXPLOSIONS?

The ignition sources can be grouped into four categories:

(a)   *Ignition by Flames and Hot Surfaces:* The flame or hot surface may have been produced by a welder or by a person bringing the open source of ignition into contact with a plant handling a combustible dust cloud.

(b)   *Spontaneous Heating:* This has already been discussed previously.

(c)   *Frictional Sparks:* Most industrial dusts are produced by mechanical action. The dust may be the main product of, for example, grinding operations, as in the case of wheat flour, or it may be an unwanted by-product, as in the case of wood dust produced during woodworking operations. In both cases, the dust is, at one point, in contact with machinery, which can be the cause of ignition. In grinders, a frequent cause of overheating is the inclusion of foreign material or tramp metal in the feed to the machine. This produces a hot spot during the grinding process, which ignites the combustible dust.

(d)   *Electrical Sources of Ignition:* Totally-enclosed motors or dust-tight apparatuses reduce the amount of dust that can penetrate into the equipment, and should be used where dust is handled. Lights should be dust-tight, so as to prevent penetration of dust into the fittings, and should be so positioned that they cannot become covered with dust, which may then begin to smoulder because of the excess heat produced by the light fittings.

WHAT ARE THE METHODS OF PREVENTING AND CONTROLLING EXPLOSIONS?

In dealing with the explosion protection necessary for the items of plants, one or more of the following methods may be used:

(a)   Avoidance of inflammable concentrations: It is necessary to avoid the formation of an inflammable concentration, therefore the concentration should be within the limits of flammability. For instance, in the pneumatic conveying field it is possible to select an air:material ratio below the explosive limit.

(b)   Enclosure of plant, processes and equipment to prevent dust escaping and reaching ignition sources.

(c)   Removal or protection of ignition sources.

(d)   Dust extraction to prevent explosible dust accumulating.

(e)   Working under inert atmosphere or under liquid.

WHAT ARE THE SECONDARY DUST EXPLOSIONS?

The worst disasters that have occurred with dust explosions have always been associated with secondary dust explosions occurring within the factory buildings. The sequence of events is usually a primary explosion in some part of the plant, which causes dust lying on beam fittings within the building to be dislodged, so producing a dust cloud of explosion concentration within the building. The dust cloud then becomes ignited and produces explosion pressure on the walls of the factory, causing their collapse. To prevent a disaster of this type, the following precautions have to be taken:

(a) *Good Housekeeping:* Regular cleaning (preferably vacuum) of the floors, walls, fittings of the workroom, etc. should be done so as to prevent the accumulation of dusts.

(b) *Exhaust Ventilation:* Dusty processes should be carried out under local exhaust ventilation, so as to prevent dust escaping into the factory buildings.

(c) *Building Design:* The design of the building is important in the prevention of secondary dust explosions. Dusty operations should be separated from non-dusty processes by partition walls of fire-resistant structures.

ARE THERE ANY SPECIAL DIFFICULTIES IN FIGHTING FIRES INVOLVING DUSTS?

Fires involving dust clouds are particularly hazardous if they are not tackled in the proper way. Fatal accidents have been caused when people have attempted to extinguish dust fires by turning a water hose on it. The water hose caused dust to be thrown into suspension in the air, and an explosion followed. Sprinkler and spray systems should be used, and jets of water should be avoided. More information about dust explosions can be found in literature, e.g. Palmer (1973).

CHECKLIST FOR CONTROLLING DANGER

1. Are we handling materials that are flammable?

2. What are the:

   (a) limits of flammability?

   (b) flashpoints?

   (c) auto-ignition temperatures?

   of the above materials.

3. What is the minimum spark ignition energy that would initiate an explosion?

4.   What is the maximum pressure that would build up in an explosion?

5.   Have all the people working in our workplace been adequately trained in fire prevention and fire protection?

Fires and explosions can be prevented or controlled in the following ways:

Prevention	*Control*
Enclosure of plants, processes and equipment to prevent flammable vapours escaping and reaching ignition sources.	Isolating and segregating the explosion. Sub-dividing the plant.
Removal or protection of ignition sources.	Venting the explosion, should it occur.
Effective extract ventilation to prevent flammable vapours accumulating.	Making the plant strong enough to withstand explosion.

REFERENCES:

1.   N. I. Sax: Dangerous Properties of Industrial Materials. (Van Nostrand Reinhold, 1979).

2.   Fire and Related Properties of Industrial Chemicals. Fire Protection Association (1974).

3.   C. J. Hilado and S. W. Sharp: Auto-ignition Temperatures of Organic Chemicals. Chemical Engineering 1972, *79* (19), 75.

4.   J. N. Bradley: Flame and Combustion Phenomena. Methuen, 1969.

5.   B. J. Lewis and J. von Elbe: Combustion, Flames and Explosion of Gases. Academic Press, 1961.

6.   D. M. Wharry and R. Hirst: Fire Technology, Chemistry and Combustion. The Institution of Fire Engineers, 1974.

7.   Manual of Firemanship, Books 1-18 (Home Office).

8.   K. N. Palmer: Dust Explosions and Fires. John Willey and Sons Inc, 1973.

PREVENTION OF DUST EXPLOSIONS
by
Milos Nedved

A dust explosion is the rapid combustion of particles of dust suspended in air. The heat produced by the combustion of these suspended particles causes rapid expansion of the surrounding gas, which in turn produces an increase in pressure at the walls of the vessel containing the dust cloud.

COMBUSTIBLE DUST

Nearly all carbonaceous materials, if they are in sufficiently fine state, can give rise to dust explosions. Also, certain metals can give rise to this penomenon, and generally metal dust explosions are much more violent than other dust explosions.

(a) *Particle Size:* Generally speaking, the finer the dust, the greater its explosibility. It means the ease with which it is ignited is enhanced, and also the rate at which pressure rises in a confined explosion increases as the particle size decreases, and similarly the maximum pressure produced increases.

(b) *Explosion Limits:* Dust clouds do exhibit the phenomena of explosion limits. In most cases, the lower explosive limit is clearly defined, whereas the upper explosive limit is not so accurately defined.

(c) *Effect of Moisture and Inert Substances:* The greater the moisture content of the dust, the lower its explosibility. However, the effect of moisture is not as great as people often assume, and we have had cases where starch dust, containing as much as 17 per cent moisture, has been involved in an explosion. An inert material present in dust will prevent an explosion. There is experimental evidence of this: for example, the maximum pressure curve of various mixtures of starch and calcium carbonate shows a gradual descent, until a limit mixture of 60 per cent calcium carbonate and 40 per cent starch was unable to be exploded.

(d) *Inert Atmospheres:* Reducing the amount of oxygen to prevent the ignition of a dust cloud is sometimes necessary in industry. Experimental tests to discover the limit at which propagation does not occur in various percentages of oxygen have shown that for the most carbonaceous dusts this figure is about 12-13 per cent, or roughly half the normal oxygen content in the air. The use of inert gas for certain industrial processes does give very effective safety, but this kind of application is usually limited to firms and processes where a large supply of inert gas is readily available.

IGNITION OF DUST CLOUDS IN INDUSTRY

(a)   *Ignition by Flames and Hot Surfaces:* The flame or hot surface may have been produced by a welder or by a person bringing an open source of ignition into contact with a plant handling a combustible dust cloud.

(b)   *Spontaneous Heating:* Certain substances, if left in layers at elevated temperatures, can start to oxidise and produce heat. If the heat produced cannot escape, the reaction accelerates and flaming combustion follows. Materials, such as wood dust and proteinic, organic materials, can give rise to this phenomena.

(c)   *Frictional Sparks:* Most industrial dusts are produced by mechanical action. The dust may be the main product of, for example, grinding operations, as in the case of wheat flour, or it may be an unwanted by-product, as in the case of wood dust produced during woodworking operations. In both cases, the dust is, at one point, in contact with machinery, which can be the cause of ignition. In grinders, a frequent cause of overheating is the inclusion of foreign material or tramp metal in the feed to the machine. This produces a hot spot during the grinding process, which ignites the combustible dust.

(d)   *Electrical Sources of Ignition:* Totally-enclosed motors of dust-tight apparatuses reduce the amount of dust that can penetrate into the equipment, and should be used where dust is handled. Lights should be dust-tight, so as to prevent penetration of dust into the fittings, and also the position of the lights should be such that they cannot become covered with dust, which may then begin to smoulder because of the excess heat produced by the lighting fittings.

EXPLOSION PROTECTION

In dealing with the explosion protection necessary for the items of plants, one or more of the following methods may be used:

(a)   *Avoidance of Inflammable Concentrations:* It is necessary to avoid the formation of an inflammable concentration, therefore the concentration should be within the limits of flammability. For instance, in the pneumatic conveying field it is possible to select an air:material ratio below the explosive limit.

(b)   *Dust Collectors:* A dust collector forms a potential hazard when it handles dust in the dry state because it always operates in the explosion region of the dust cloud. If possible, it is best to substitute dry dust collectors by wet, scrubber-type collectors. Wet scrubber dust collectors effectively water-drench any suspended dust cloud, and so eliminate the hazard.

48

(c)   *Explosion Relief:*   The most common method of protecting industrial plants against the effects of a dust explosion is to provide an area of weakness in its structure.   In this way, should an explosion develop in the plant, the area of weakness will allow the expanded gases to escape, thus reducing the pressure produced by the explosion.

(d)   *Explosion Vent Covers:*   An open explosion relief vent is, by far, the most effective method of preventing excessive pressure development in an industrial plant.   However, it is not always practicable, under industrial conditions, to have an open relief in the plant: generally the relief has to be covered in some way or other.   All relief covers reduce the efficiency of a vent, to some extent, but experiments have shown that certain types of relief covers do not interfere to a very great extent, e.g. bursting panels.   In this type of a vent, a panel of thin material is placed across the vent opening.   If an explosion develops, the panel bursts open and allows the combustion products to escape.

(e)   *Explosion Suppression:*   In recent years, a system of explosion suppression has been developed which depends upon the detection of the early stages of a dust explosion.   The detector, which is based on the anaeroid barometer principle, is sensitive to an increase of pressure above the ambient pressure of the system. The detector is coupled to suppression cells which contain a suppressant such as chlorobromomethane.   The cells are situated at various points in the industrial plant, and as soon as the detector detects an explosion commencing, the suppressant is discharged into the plant by means of detonators situated inside the cells.   In this way, it is possible to protect a plant from the effects of a dust explosion when other conventional means are not satisfactory.

SECONDARY EXPLOSIONS IN FACTORIES

The worst disasters that have occurred with dust explosions have always been associated with a secondary dust explosion occurring within the factory buildings. The sequence of events is usually a primary explosion in some part of the plant, which causes dust lying on beam fittings within the building to be dislodged, and so to produce a dust cloud of explosive concentration within the building.   The dust cloud then becomes ignited and produces explosion pressure on the walls of the factory, causing their collapse.   To prevent a disaster of this type, the following precautions have to be taken:

(a)   *Good Housekeeping:*   Regular cleaning (preferably vacuum) of the floors, walls, fittings of the workroom, etc. so as to prevent the accumulation of dusts.

(b)   *Exhaust Ventilation:*   Dusty processes should be carried out under local exhaust ventilation, so as to prevent dust escaping into the factory buildings.

(c)   *Building Design:* The design of the building is important in the prevention of secondary dust explosions. Dusty operations should be separated from non-dusty processes by partition walls of fire-resistant construction.

FIRES INVOLVING FLAMMABLE DUSTS

Fires involving dust clouds are particularly hazardous if they are not tackled in a proper way. Fatal accidents have been caused when people have attempted to extinguish a dust fire by turning a water hose on it. The water hose caused dust to be thrown into suspension in the air, and an explosion followed. Sprinkler and spray systems should be used, but jets of water should be avoided.

APPENDICES:

Appendix   I.   Explosion and Fires in Plant and/or Buildings Causing Very Extensive Damage and/or Fatalities.

Appendix   II. List of Dusts Tested for Explosibility in the Form of Dust Cloud.

Appendix   III. Materials Recently Involved in Industrial Dust Explosions, With Serious Consequences.

EXPLOSIONS AND FIRES IN PLANT AND/OR BUILDINGS CAUSING VERY EXTENSIVE DAMAGE AND/OR FATALITIES

SUBSTANCES	*PROCESSES, PLANT AND PREMISES INVOLVED IN INCIDENT*
Aluminium	Blown powder production: settling chambers, cyclones, bag filter units, conveyors and buildings. Stamped powder production: stamped machines, ducting, collecting plant and workrooms. Milled powder production: ball mills Metal spraying: dust collectors and workrooms.
Aluminium stearate	Milling machines, pneumatic dryers and bag filter units.
Calico and metal dusts from polishing processes	Bag filter units (one building severely damaged by fire).
Cattle cake Compound cake Cotton seed Feedstuff Provender Sunflower seed cake Linseed	Milling machines, disintegrators, bucket elevators, cyclones and bag filter units: Rotary drum dryers, storage bins, silos and workrooms.
Coal	Milling machines, classifiers, cyclones, ducting and bag filter units in the production of pulverised fuel.
Cork	Milling machines, conveyors, screens, cyclones, bag filter units and workrooms
Dextrine and core gum	Kilns, blending plant, humidifiers and disintegrators.
Esparto grass	Deduster machines, cyclones and bag filters.
Ferromanganese	Milling, pneumatic conveying and screening plant.

SUBSTANCES	PROCESSES, PLANT AND PREMISES INVOLVED IN INCIDENT
Gramophone records	Milling machines and cyclones.
Magnesium	Hammer mills, ball mills, pan mills, conveyors, screens, cyclones and settling chambers (including improperly maintained scrubbers) associated with pulverising, abrading and scratching processes.
Maize husk	Grinding plant and building.
Malt	Mill and bucket elevators in distilleries (less serious explosions in breweries).
Polystyrene	Extruding machines, pneumatic dryers, cyclones, bag filter units, storage bins and workrooms.
Rubber	Dust chambers and collecting plant associated with abrasive machines used for tyre treading and retreading operations.
Starch	Starch powder production: disintegrators, cyclones, bag filter units, bucket elevators and pneumatic dryers. Starch moulding of sugar confectionery: workroom.
Wood flour	Milling plant, cyclones, screens and storage bins, associated with pulverising operations.

LIST OF DUSTS TESTED FOR EXPLOSIBILITY
IN THE FORM OF DUST CLOUD

Acacia gum
ParaAcetaminobenzene
  sulphonyl chloride
Acetanilide
Acetoacetanilide
Acetoparaphenetidine
Acetyl para-nitro ortho-
  toluidine
Acetyl sulphonamide
Acrylamide/acrylic acid
  copolymer
Acrylic polymer
Acrylonitrile polymer
Acrylonitrile vinylidene
  chloride copolymer
Albertol synthetic gum
Albumen
Alginate powder
Alizarine, black R
Alkyd powder coatings
Alkylnapthalenesulphonates
Aluminium metallic dust
Aluminium octoate
Aluminium swarf
2-Aminothiazole
Anthracite (ground)
Anthraquinone
Aspirin
Azo compounds, various
Azodicarbonamide

Balsa wood
Barium stearate
Battery case mix (coal
  dust, rubber and
  synthetic resins)

Beet sugar pulp
Benzoic acid
Benzophenone hydrazone
Benzoyl peroxide
  (with 80 per cent inert)
Benzthlazole disulphide
Biscuit dust
Bitumen
Blood, dried
Bone flour, steamed
Bone meal
Bran grit
Brazil nut
Bread crumbs
Briquette coal
Bronze powder

Cadmium yellow
Calcium pentothenate
Calcium propionate
Calcium silicide
Calcium stearate
Calt meals
Caprolactam monomer
2-Carbamoy 1-oxymethyl-
  1-methyl-5nitrolmidazole
Carboxyethylmethyl
  cellulose
Carnauba wax
Carob powder
Carrot dust, dried
Cascara sagrada
Casein
Casein formaldehyde resin
Cashew polymer resin
Cattle food, various

Cellulose
Cellulose acetate butyrate
Cellulose triacetate
Charcoal wood (ball-milled)
Chicory
Chloro-amino toluene
  sulphonic acid
Para-Chloro-ortho toluidine
  hydrochloride
Chocolate
Chocolate, milk chocolate
  crumb
4-Chloro 2:5 diethoxynitro-
  benzene
4-Chloro 2:5 dimethoxy
  nitro-benzene
Chrome leather residues
Chrysoidine
Chinchona bark
Citric acid
Coconut shell dust
Coffee, finely ground
  spray dried extract
Coffee bean husk
Coffee, instant
Coffee, pulverised
  ground
Colt, ground fibre
Copal resin
Core gum
Cork/rubber composi-
  tions
Cortisone acetate
Cotton
  ground
  chopped
  fly

Cotton seed hull
  'pepper'
Crotonic acid/vinyl
  acetate copolymer
Cuprous cyanide
Cyclohexanone peroxide
  (with filter)
Cychlonexyl ammonium
  cyclamate
N-Cyclohexylthio
  phthalimide

Detergents, various
Dextrose monohydrate
Diazo compounds
Dibutyl tin malleate
Dibutyl tin oxide
Dicyclohexyl phthalate
Dihydrostreptomycin,
  crystalline
Dihydrostreptomycinsul-
  phate
P-(D1-2 hydroxyethoxy)
  benzene
Dimethyl acridan
Dimethyl diphenyl urea
Dinitroaniline
Dinitro stilbene disul-
  phonic acid
3:5 Dinitrobenzoyl-
  chloride
2-2:4 Dinitrophenyl
  thiobenzthiuzole
Diotcyl tin oxide
Diphenyl guanidine
Diphenylol propane
Distillery meal
1:5 Disulpho anthrazuinone
4:4 Dithiomorpholine

Ebonite
Ephedra herb
Epoxy resin dust
Erqot of rye

Ethyl cellulose
Ethylene thiourea

Food, animal (general)
Fermentation residues
Ferrochrome
Ferrosilicon (75% Si)
Ferrotantalum niobium
Ferrotitanium
Flax dust
Flax and jute mixture
Flour/fat mixtures
Fragula bark
Fungicides (various, organic)

Galto gum
Gelatin
Gentian root
Glass reinforced polyester
  (G.R.P.)
Glue, ground animal
Gluten,
  wheat
  meal
Glyceryl monostearate
Grass, dried
Gums
  Acacia
  Arabic
  British
  Core
  Gatto
  Karaya
  Kauri
  Rachig
  Synthetic
  Tragacanth
  Yacca

Hardboard dust
Heptabarbitone
Herbs, ground mixed
Hoof and horn (hydrolysed)
Hops, spent

Horn,
  Dust
Meal
Hydrazides
Hydrocortisone acetate
Hydrocortisone alcohol
Hydroquinone
Hydroxyethyl cellulose
Hydroxyethyl methyl
  cellulose

Ipecacuanha root
Iron, metallic:
  carbonyl iron powder
  electrolytic powder
  metallurgical powder
N-IsopropylN-phenyl para-
  phenylene diamine

Jaborandi leaf
Jalap root

Karaya gum
Kauri gum

Lactose
Lauryl peroxide
Lead refinery dust
Lead stearate
Leather, ground
Leather/rubber composition
Linoleum
Lint
Liquorice root
Lithium hydroxystearate
Liver extract
Locust bean kernel
Locust meal

Magnesium,
  swarf
Magnesium stearate
Male fern

Manganese ethylene bis
  dithiocarbamate
Manioc
Melamine resins
2:Mercaptobenzthiozole
Methyl methacrylate
Milk powder
Mimosa bark
Monochloracetic acid
Mould bran
Mustard offal

Beta Naphthol
Nigrosine Hydro-
  chloride
para-Nitro-ortho-
  anisidine

para-Nitro benzoic acid
Nitrocellulose
2-Nitro-diphenylamine
Nitrofurazone
Nitrofurfural semincarba-
  zone
Nitrotoluidines
Nut shell, crushed
Nutmeg, ground
Nylon flock
Nylon II

Oatmeal
Obechi wood
Oil cake

Palm kernel cake

Pancreatin
Paraffin wax
Paste, powder (builder's
Pea flour
Peat
Pectin
Penicillin, N-ethyl
  piperadine salt
Penicillin, sodium salt
Phenol-formaldehyde
  resin
Phenothiazine
Phenyl carbamates
  (various)
Photographic powder
Plumstone dust
Polyester resin

MATERIALS RECENTLY INVOLVED IN INDUSTRIAL DUST EXPLOSIONS WITH SERIOUS CONSEQUENCES

Apidic acid
Aluminium
Atomised powder
Dust from metal spraying
   process
Dust from finishing process
Dust from polishing process
Aluminium laurate
Aluminium stearate
Animal feedstuff
Anthracene
Asbestos (resinated)

Barley milled
Battery case grindings

Cattle cake (mainly linseed)
Celluloid
Cellulose acetate
Cellulose nitrate
Cereals ground
Chicken manure
Coal
Cocoa
Cocoa bean husk
Compound cake
Cork
Corn flour

Dextrin
Distillery grain solubles
Dyestuffs, various

Esparto grass

Ferromanganese
Fish meal
Flour (All cereals)

Glucose
Grains
Ground nut meal, extracted

Hexamine

Linseed

Magnesium,
   powder
Maize
Maize husk
Maize starch
Malt
Meat meal
Methyl cellulose

Naphthalene

Oak husk

Paper dust
Paraformaldehyde
Pentaerythritol
Pepper, pesticides,
   various
Phenolic resin mixtures
   and moulding powders
Phosphorus, red
Phthalic anhydride

RECENT DEVELOPMENTS IN MAJOR HAZARD CONTROL METHODS

by
Milos Nedved

IDENTIFICATION OF MAJOR HAZARDS IN INDUSTRIAL PROCESSES

Identification Criteria

A "major hazard" is defined as any industrial activity using or producing dangerous substances in such a quantity that they possess the potential to cause extensive damage and kill or injure persons within or outside the site boundary. The term is also intended to cover the transportation of dangerous substances, whether by road, rail or sea, during loading and discharge at installations.

The first step in any major hazard control system is to establish suitable criteria by which to recognise those installations presenting the greatest potential threat to safety.

While there is a large range of events which could cause an emergency situation, the following types of materials would generally be involved:

(a)   toxic gases which, following release in tonnage quantities, are lethal or harmful at considerable distances from the point of release;

(b)   extremely toxic material which, following release and dispersion in kilogramme quantities, is lethal or harmful for considerable distances from the point of release;

(c)   flammable liquids or gases which, following release in tonnage quantities, form a large flammable cloud, which in turn burns or explodes; and

(d)   unstable or highly reactive materials which have exploded.

Notification Requirements

Once the major hazard criteria have been identified, notification of the hazard to the relevant health and safety enforcement authority is absolutely fundamental to any system of control. Attention can then be given to ensuring that existing and proposed installations are made as safe as possible, and where efforts to prevent a major accident fail, that the consequences affect as few people as possible.

If, therefore, an installation has, or is liable to have, present a dangerous substance identified under the criteria discussed earlier in a quantity greater than the threshold level, a formal notification should be made. The notification should contain the following information:

(a)   name and address of the notifier;

(b)   address of installation and its map reference;

(c)   names and maximum quantities of the dangerous substances liable to be at the installation;

(d)   general description of the activities carried on;

(e)   date on which a new activity at installation is proposed to start; and

(f)   name and address of any relevant planning authority.

Accident/Incident Reporting

An important feature of any control system is adequate reporting of accidents and "near misses" involving dangerous substances. Full investigations and proper analyses can then be made and recurrences can be prevented, and faults in plant and systems identified and corrected so as to enable lessons to be drawn from experience.

A "major accident" is defined as an occurrence (including, in particular, a major emission, fire or explosion) resulting from *uncontrolled developments* in the course of an industrial activity, leading to *serious danger to persons*, whether immediate or delayed, inside or outside the installation, or to the environment, and involving one or more dangerous substances. The term "uncontrolled developments" should be taken to mean that the occurrences of concern are likely to develop quickly, to be outside the normal, expected range of operating problems, to present only limited opportunities for preventive action and to require any such action to be in the nature of an emergency response.

"Serious danger to persons" should be taken to mean death or serious injury, including injury to health or the threat of death or serious injury, whether caused immediately or as a delayed effect.

Whenever such an incident occurs, the person having control of the installation must report it forthwith to the relevant health and safety enforcement authority. The report should contain the following types information:

(a)   circumstances of the incident;

(b)   dangerous substances involved;

(c)   data available for assessing the effect of the incident on persons;

(d)   emergency measures taken;

(e)   steps to be taken to alleviate any medium or long-term effects; and

(f)   steps to be taken to prevent a recurrence.

58

EMERGENCY OPERATIONS

"Major hazard installations", by definition, contain residual hazards, even when properly protected, but substantial mitigating measures can be taken to reduce their impact on people outside the installation.

Firstly, the siting of the installation is important. It should be located away from centres of population, and substantial growth of population around it should be avoided. Experience has shown that there is a tendency, mainly in developing countries, for workers and their families to set up dwellings adjacent to the installation boundary. Secondly, the effects of any major accident that might occur can be minimised by putting into effect emergency plans covering both the installation and the neighbourhood.

Siting Policies

Separation between a major hazard installation and the nearest communities is a prudent measure. For a new installation on a "green-field" site, it is a straightforward matter, in that the installation can be located well away from the general public. Control, then, needs to be exerted to stop the public and other vulnerable developments from encroaching on the site. For existing installations, the difficulty is greater. The separation distance does not provide any extra protection for the workers or the environment. Generally, the level of risk from a well-run installation would not justify imposing a separation distance at the expense of either reducing the activity at the installation or removing nearby buildings. However, planning controls will need to be exerted to prevent such situations from developing or worsening. The separation distances should be based on a quantitative assessment of the consequences that could occur as a result of a major accident.

Consequence Criteria

There is a large range of major accidents that could cause emergency situations, but those which could cause injury to a neighbouring population are limited in number, and probably fall within the following categories:

Events Involving Flammable Materials

(a)   Major fires with no explosion danger: hazards from prolonged high levels of thermal radiation and smoke.

(b)   Fire threatening items of plant containing hazardous substances: hazards from spread of fire, explosion or release of toxic substances.

(c)   Explosion with little or no warning: hazards from blast wave, flying debris and high levels of thermal radiation.

Events Involving Toxic Materials

(a)   Slow or intermittent release of toxic substances, e.g. through leaking valve.

(b)   Items of plant containing toxic substances threatened by fire: hazards from potential loss of containment.

(c)   Rapid release of limited duration, due to plant failure, e.g. fracture of pipe: hazards from toxic cloud, limited in size, which may quickly disperse.

(d)   Massive release of toxic substances, due to failure of large storage or process vessel, or uncontrollable chemical reaction and failure of safety systems: exposure hazard would affect a wide area.

The major difference between the release of toxic and flammable materials is that toxic materials tend to be hazardous over greater distances. A toxic cloud can travel with the wind for some distance, whereas a flammable cloud is likely to break up, find a source of ignition or be diluted to below the lower flammable limit (LFL) over a relatively short distance.

Considerable information is available on possible events involving LPG based on historical data, field trials and computer modelling. The consequences arising from release of LPG are fire and/or explosion. Larger events would either be in the form of a boiling liquid expanding vapour explosion (BLEVE) or an unconfined vapour cloud explosion (UVCE). Smaller releases from pipework could also have off-site consequences. In major incidents, a UVCE may occur without warning as a result of a vessel failure. A BLEVE produces high levels of thermal radiation (Mexico City disaster).

The hazard consequences presented by highly flammable liquids at ambient conditions (flashpoint less than $21^{\circ}C$) are much less than those of LPG, as the only potential off-site problem is that of thermal radiation. Even the largest installations, containing 10,000 tonnes or more, should not pose any significant problems beyond a distance of 250 metres.

The consequences of toxic releases are time-dependent and will vary with distance and prevailing weather conditions from the point of release. To derive hazard ranges, therefore, it is necessary to be able to estimate the concentrations and durations of the gas clouds at various distances downwind of the release point. This information can then be combined with a definition of the hazardous dose, to deduce the maximum range.

Development Controls

First of all, a zone should be established around the major hazard installation at a distance determined according to the consequence criteria. Outside the zone, a properly run installation should not present any serious consequences in the event of an emergency, with the possible exception of a large-scale release of a toxic gas which may be present in a dangerous concentration for some distance downwind beyond the periphery of the zone.

As a guide to the size of zone envisaged where LPG is the hazard, it would vary from 300-1,500 metres around the installation, depending on the latter's inventory. For flammable gases, the corresponding figure would be about 500 metres; for toxic gases (e.g. chlorine), 1,000-1,500 metres; for highly reactive substances (e.g. organic peroxides), 250 metres; and for highly flammable liquids, 250 metres.

Next, the proposed development within the zones can be the subject of consultation between the relevant authorities to assess whether the residual hazard presented by the major hazard installation poses an unacceptable threat to the proposed development and, thus, whether or not to allow the development.

Here, the following factors are important:

(a)   proportion of time spent by particular individuals in the type of development (compare homes, factories, shops, hotels, etc.);

(b)   size of the development in terms of number of users;

(c)   ease of evacuation or other emergency measures;

(d)   vulnerability of the population (compare children, disabled, elderly, factory workers); and

(e)   physical factors (height of buildings, type of construction, terrain, etc.).

Generally speaking, there should be no reason for wishing to prevent any development beyond the zones on health and safety grounds. On the other hand, any development which would contain a large population, or particularly vulnerable populations, e.g. children in schools or hospital patients, should not be sited within the zones. For other cases, the above criteria would require full consideration as part of the judgement.

The situation can also work in reverse, i.e. for proposed increases in capacity at a major hazard installation. The use of the consequence criteria, to modify the zone as the residual hazard increases, can help to assess any increased threat to the community and, thus, whether to allow the development.

Emergency Planning

Emergency planning is just one means of ensuring safety. It cannot be considered in isolation, and its proper place is as a backup to the preventive measures discussed earlier in this paper. The employer bears most of the responsibility for these measures, which can be summarised as follows:

(a)   ensure that the plant and storage vessels are designed and installed to a good standard, and in a location where hazards are unlikely to arise from other plant or buildings nearby;

(b)   ensure that proper work routines and effective maintenance procedures are set up;

(c)   assess what could still happen to cause an emergency situation: further preventive measures may still be possible;

(d)   assess what dangers could arise to people, both on-site and off-site, as a result of these foreseeable emergencies, and how these could be mitigated by pre-planned remedial and rescue measures, utilising the resources of the major hazard installations and the public emergency services.

Item (d) will require the formulation of an emergency plan. This will need to be in two separate parts: the on-site plan, covering the action to be taken on the installation, and the off-site plan, covering the action to be taken in the neighbourhood around it. These are considered separately.

Formulation of On-Site Emergency Plan

Employers should set up a plan, based on the specific needs of the particular site, for dealing with emergencies which may occur even after all precautions have been taken.

Useful guidance on the preparation of the plan is contained in the Chemical Industries Association booklet "Recommended Procedures for Handling Major Emergencies" (U.K.).

Each plan would be an individual document for each installation, reflecting the particular hazards associated with its plant, processes and location. All plans would, however, need to contain the following key elements:

(a)   assessment of the size and nature of the events foreseen, and the probability that they might occur;

(b)   formulation of the plan and liaison with outside authorities, including emergency services;

(c)   setting up of emergency control centres;

(d)   action on-site: fire-fighting procedures, rescue systems, evacuation arrangements, first-aid arrangements; and

(e)   plant shut-down procedures.

Formulation of Off-Site Emergency Plan

The responsibility for setting up this part of the plan should lie with the public authority that has control over the resources and facilities in the neighbourhood of the unstallation. While the formulation of the off-site plan will be the responsibility of the relevant public authorities, the employer will possess detailed knowledge about his plant and off-site plan. This includes information on hazardous substances stored and used, with details of their physical and chemical properties; identification of foreseeable major accidents with off-site consequences; evaluation of consequence criteria to enable the zones to be established to form the basis of planning control and emergency planning; and details of the on-site emergency plan. Thus, the off-site and on-site plans should complement each other.

During the setting up of the plan, there should be full liaison between all the parties involved, including the employer and the local authority. Where relevant, the following should also be involved: fire authorities, water authorities, health authorities, ambulance services, gas and electricity suppliers, police, railways, Government departments and agencies.

Finally, no matter how well the plans are drafted, experience has shown that in real emergencies there is often much confusion. It is essential, therefore, that plans are regularly rehearsed and practised under simulated emergency conditions in order to test the response of operators and emergency service personnel, test and evaluate the interaction of the various organisations involved, evaluate the effectiveness of the equipment, and increase gains from experience.

TRAINING ACTIVITIES

While provision for the training of managers, supervisors and workers in plant operations is often included in arrangements for the transfer of technology and industrial processes, this is not always the case as regards measures to ensure high safety levels. Even in instances where safety provisions are included in technology transfer packages, the limitations of the safety and health institutions in the receiving country, as well as the lack of general awareness of safety and health problems, have been identified as major obstacles in the way of overcoming the negative effects of transfer of technology. A prerequisite for ensuring high levels of safety and health during the transfer of modern technology is the existence of a national infrastructure of institutions and systems to cope with all aspects of technology transfer related to safety and health. The following are suggested as discussion points for some of the more specific and important training needs for major hazard controls:

Workers

(a)   Relevant safety and health provisions associated with the hazardous substances stored and processed.

(b)   Correct operating procedures.

(c)   Safe systems of work.

(d)   Plant emergency procedures.

Employers

Supervisors

(a)   Relevant safety and health provisions associated with hazardous substances stored and processed.

(b)   Role of the supervisor for ensuring safety and health.

(c)   Monitoring correct operating procedures.

(d)   Monitoring safe systems of work.

(e)   Plant emergency procedures.

Senior Management

(a)   Hazard assessment techniques.

(b)   Development of consequence criteria.

(c)   Emergency planning.

Local Government

(a)   Off-site emergency planning.

(b)   Emergency operations (fire-fighting, evacuation, rescue, first-aid, etc.)

(c)   Siting policies, use of consequence criteria development controls.

EXCHANGE OF INFORMATION

Detailed information about incidents that have occurred, such as what went wrong, why it went wrong, what happened next, details of casualties, consequences, mitigating measures, etc., can play an invaluable role in preventing a recurrence of such incidents.

The ILO International Occupational Safety and Health Information Centre (CIS) actively disseminates information through its network of nearly 50 national centres. Its publications (CIS Abstracts and Indices) are entered in the CIS data base, which is available on-line.

The ILO also operates the International Occupational Safety Health Hazard Alert System. The system is designed to assist countries in the exchange of information on occupational safety and health hazards, and their prevention. After an accident occurs, the system can disseminate, to its 100 member countries, reports prepared by the national authorities, to help other countries avoid similar catastrophes.

64

COMPUTER-BASED INFORMATION SYSTEMS

The main advantage of computer-based information retrieval systems is that they allow large data bases to be searched quickly, and retrieve and list those items which are of interest to the user. The information may be stored in either textual or coded form, or as a combination of both, using "key words" to assist in the retrieval of information. Within the context of the control of major hazards, there are already in existence some data bases containing relevant information of international incidents, of which the following are examples:

(a)   "FACTS". This data bank is operated by TNO based in Apeldoorn, The Netherlands, and contains technical data of incidents with dangerous materials that happened during storage, transhipment, transport, use and disposal of chemicals. It also includes information on near-misses. At the end of 1982, the data bank contained information on more than 6,000 incidents, the majority of which had occurred during the previous five years. These incidents cover approximately 450 chemicals, the most commonly recorded being those involving fuel oil, followed by LPG and natural gas. Of the toxic materials, chlorine features most often, with approximately 250 recorded incidents, followed by ammonia, with approximately 150 incidents.

(b)   "CHAFINCH". This is the data bank of chemical accident failure incidents and chemical hazards operated by Risk Management Ltd., and is available on-line throughout Europe, via SIA Computer Services. The data bank was last updated in March 1982, after which between 1,000 and 1,500 incidents were on record, categorised under 47 headings.

Based on the ILO Working Paper on Control of Major Hazards in Industry and Prevention of Major Accidents (Geneva 1985).

TECHNIQUES OF INSPECTION OF MAJOR HAZARD WORKS
by
Milos Nedved

The majority of serious incidents arise from the loss of containment of hazardous substances. Therefore, it is necessary, in the first instance, to identify those plant items from which, in the event of failure, a quantity of hazardous substances which would prejudice the safety of the adjacent population and plant personnel could be released. Local circumstances will influence the selection of plant items for closer scrutiny. When such items of plant have been identified, the sampling techniques described below should be followed.

Labour inspectors should be aware, in relation to the plant they intend to examine, of the major incidents which have occurred worldwide. The lessons learned from such incidents should be implemented to prevent a repetition of such mistakes. Containment may be achieved by:

(a)   establishing the activity on a sound basis;

(b)   establishing and maintaining plant integrity; and

(c)   controlling process deviations.

STANDARD CONDITIONS OF PROCESS, STORAGE, ETC.

At the outset, inspectors should be satisfied:

(a)   that they are aware of all the activities, i.e. process and storage that are carried out with respect to the plant in question;

(b)   with the steps taken by the occupier to ensure the safety and integrity of these activities. A technical appraisal should be made of the occupier's competence and his system for controlling the competence of contractors;

(c)   that senior staff responsible for the plant thoroughly understand the process and reaction conditions, and their implications. Matters for discussion will include:

(i)   choice of continuous or batch process,

(ii)   choice of inventory catalyst, and

(iii)   operating temperature, pressure, residence time;

(d)   that specialist inspectors can advise on the suitability of the choices made; and

(e)   with the management control systems.

The process documentation should be scrutinised for completeness and quality. It should preferably be compiled at several levels and aimed at different users, i.e. process description, operating instructions, quality control manuals, safety information sheets, emergency procedures. The nature of such documents depends upon the operations, from the simple case of bulk LPG storage, for use as a fuel, to complex chemical manufacturing plants.

PROCESS DESCRIPTION

This should record both basic and detailed knowledge, and include:

(a)   principles involved and process rationale;

(b)   basic chemistry and physics of the main reaction, and of any side reactions;

(c)   hazards of all materials involved and safety precautions;

(d)   description of the plant, equipment list and line diagram;

(e)   arrangements for transport, storage and transfer of raw materials;

(f)   detailed processing conditions, including the allowable limits for each operating parameter;

(g)   arrangements for keeping within these limits;

(h)   arrangements for transfer, storage and transport of finished products;

(i)   specification of the operating procedures; and

(j)   procedure for authorising process changes.

OPERATING INSTRUCTIONS

They should be consistent with the process description. There should be operating instructions for each plant and storage facility, which should:

(a)   be in a form that can be clearly understood by the operators;

(b)   describe the plant following the sequence of the flow of work;

(c)   clearly state the conditions to be maintained;

(d)   explain the action of the various process controls and monitoring instruments;

(e)   specify allowable excursion limits, and the action to be taken in the event of excursion limits being reached;

(f)   specify emergency procedures, including the reporting route; and

(g)   specify procedures for initial and refresher training of operators.

All potentially hazardous processes should be monitored, and records should be kept, in particular, of excursions beyond the safe operating limits.

Inspectors should be satisfied that the firm has sufficient knowledge of the materials and processes, and that the normal process arrangements have been adequately defined. It is particularly important to establish that the operators will seek help at the appropriate time. Where the plant is computer-controlled, special considerations apply. The inspector will check that there are suitable procedures for keeping within the defined standard conditions. The inspector will normally assess the selection and training of operators.

PLANT INTEGRITY

To establish and maintain the mechanical integrity of pressure systems and associated plant, the following principles should be adopted:

The safe working limits of the plant should be carefully established for use as design data, and should be compatible with the normal process operating conditions.

In addition to the range of pressures, temperatures and loadings considered, the design data established should encompass relevant information on flow rates, operation cycles, basis for pressure relief (liquid, gaseous or explosion) and properties of substances to be contained.

The plant should be designed, constructed, inspected, tested and certified, as necessary, to an appropriate design and construction code which will achieve the safe working limits.

Correct design is paramount to the safety of pressure vessels and associated plant, and is reliant on the design data provided. Incorrect design could lead to all the known modes of failure. Selection of wrong materials may contribute to stress corrosion cracking or brittle fracture. Failure to allow for high temperature use, cyclic operations, vacuum conditions, correct design strengths and corrosion may initiate failures due to creep, fatigue, buckling and overstress.

The appropriate design and construction code chosen for the manufacture of a pressure vessel should be applied in its entirety, and complementary use of other codes should only be contemplated where the chosen code is deficient. (Use of more than one code to achieve economic manufacture *may* result in an unsafe vessel, and the certification should be closely checked).

It should be noted that design to a pressure vessel code, such as British Standard 5500, does not necessarily indicate the design to be correct. Such codes contain optional rules for features such as low temperature use, fatigue, vacuum operation and corrosion allowance. Requirements for a particular vessel should be checked against the vessel certification. Certain vessels may also have specified design lives.

The manufacturing, inspection and testing requirements will be specified in the code, which will also list the vessel particulars to be included on the certification, stamping or nameplates. If a particular requirement for a vessel, such as ability to support vacuum of low temperature use, is not marked on the vessel or its certification, its ability to operate safely in these conditions should not be assumed.

If it is foreseeable that in abnormal conditions excursions outside the safe working limits of the plant may occur, protective devices should be fitted or safe systems of work devised to maintain plant integrity.

Protective devices may include pressure/vacuum relief valves, bursting discs, explosion vents, high/low level alarms, high/low temperature alarms, weight tables, contents gauges, maximum fill devices, padding air/gas reducing valve stations and many forms of instrumentation. Such devices should be carefully checked as being suitable for the purpose intended. For example, a bursting disc of the wrong design or wrongly fitted may not relieve at the maximum working pressure of the vessel; padding arrangements, in conjunction with fluids subject to pressure increase with temperature, may promote overpressure conditions, if not properly regulated.

Safe systems of work may include manual venting procedures, loading/unloading systems and temperature control at fill. Cold contents stored in vessels at temperatures below their minimum safe working temperatures may promote brittle failure.

Protective devices only operate infrequently, usually when process control has failed. It is important, therefore, that reasonable measures are taken to ensure that they function when required. A schedule of protective devices should be prepared and adequate inspection/testing schemes devised. The frequency and scope of such inspection/testing should be decided on by a competent person with due regard to local conditions, such as environment, consequences of failure, duplication, fail-safe provision and risk of physical damage.

The plant should be properly maintained. Plant maintenance should be acceptable, and standards determined and overseen by a competent engineer of appropriate discipline. Where necessary, planned preventive or condition-based maintenance schemes should be established. The objective should be to maintain plant integrity at the established safe working limits. Where this is not achieved, the plant should be down-rated to the repair standard and protective devices modified accordingly. With pressure systems, care should be taken to extend maintenance procedures to include essential associated plant, such as refrigerating systems and reducing valve stations on padding air/gas supplies.

The plant should be periodically and appropriately inspected and/or tested. The periodicity and scope of plant inspections, and the inspection techniques to be used, should be decided by a competent inspection authority and should reflect the particular hazards and vulnerability of the contents stored or processed, and the

70

storage or process plant. The inspection techniques used should be seen to be suitable for the particular application. Inspections of pressure vessels should extend to support structures, holding down bolts and foundations, where applicable.

Any significant deterioration found should be recorded in the inspection report, with particulars of the inspection techniques used. The vessel should be reassessed for continued use with the particular contents, and the safe working limits confirmed or revised. Any significant weld or material defects found should be critically examined with regard to effects on vessel integrity, origin and likelihood of recurrence. Where defects are ground out or rewelded, the effects of safe working pressure should be carefully analysed. Any stress analysis should pay due regard to residual and operational stresses, with appropriate stress intensification factors, where these apply. Where defects are left or are likely to recur, their significance should be assessed by a fracture mechanics analysis, and should be suitably monitored when the vessel is returned to use. Where any metal removal or thinning has occurred, due consideration must be given to safe working pressure limits based on the remaining metal thickness. Certification should specify, as appropriate:

(a)   maximum/minimum safe working pressure, particularly the ability to support partial or full vacuum where vacuum conditions can be foreseen;

(b)   maximum/minimum safe working temperature, particularly where contents may be at sub-zero temperature;

(c)   maximum permissible load for the vessel and its supports, which becomes important when significant deterioration has occurred in the vessel or supports/ foundations, where a change of contents is intended, or when a site hydraulic test is proposed;

(d)   fill ratio, which determines the maximum fill permitted in vessels containing liquefiable gases, where the contents may expand with foreseeable temperature rise. It is expressed as the ratio of the mass of contents introduced to the mass of water which effectively fills the storage vessel. Care should be taken that the maximum fill permitted to avoid liquid locking is compatible with the maximum permissible load for the vessel and its supports; and

(e)   date of next examination.

EVALUATION OF PLANT INTEGRITY

The following guidelines may be generally useful in assessing the integrity of pressure systems with hazardous contents and the effectiveness of management control:

Vessels

Spherical and cylindrical pressure vessels should have been examined by a competent inspection authority within the last 5 years. Flat-bottomed storage vessels should have been examined within the last 10 years. Vessels with known defects should be examined on a shorter time-scale. The vessels should have known safe working limits of pressure, temperature and loading, certified by the competent inspection authority. Protective devices should be fitted or safe systems of work devised to prevent any foreseeable excursion outside the safe working limits. Such devices should be appropriately inspected or tested. Pressure/vacuum relief devices should be examined within a period specified by the makers or a competent person, but not exceeding 2 years. The date of the next vessel inspection should be certified by the competent inspection authority.

Pipelines

Pipelines containing hazardous fluids should be designed and contructed to an appropriate standard, such as British Standard 3351. Particular causes of failure are inadequate allowance for expansion/contraction, inadequate supports, metal thinning due to corrosion/erosion, poorly-made joints and inadequate inspection. The safe working limits of pressure, temperature and loading should be known, and protective devices fitted where excursion is foreseeable. Any modifications or extensions made should be of a standard at least equivalent to the original design and construction code, and overseen by a competent person. A planned inspection procedure should be followed and the scope and periodicity determined by a competent person, and appropriate to the foreseeable defects and hazards associated with the pipeline contents or environment. There should be a reliable system for unambiguous identification of pipelines. The pitfalls of any particular system, e.g. colour coding, should have been considered.

Pumps and Ancillary Equipment

Inspection and maintenance of pumps and ancillary equipment are only of consequence if failure could create a hazardous situation by loss of fluids or loss of pressure, rather than nuisance interruption to the process. Where failure can be hazardous, standby equipment should be provided. If standby equipment is not provided, planned or preventive maintenance systems should be followed. The design parameters of the pump and equipment should be known (by maker's or competent person's certification) and shown to be suitable for the particular fluid being processed and the duty demanded. Emergency procedures should be available in the event of pump or equipment failure.

Modifications to Plant and Equipment

Modifications to plant and equipment should be carried out to a standard at least equivalent to the original integrity, or the plant down-rated accordingly. On completion, the modifications should be inspected and/or tested by a competent inspection authority and approved for use, and details recorded. The modifications should be subjected to at least the same inspection and maintenance procedures as the remaining plant.

Management Control

Any evaluation of management control to establish and maintain plant integrity should be considered in conjunction with the area inspector with regard to the foregoing. Its effectiveness may be demonstrated by the ability to produce ready evidence of pertinent design and construction codes, industrial codes, project management, planned or preventive maintenance systems and periodic inspection and assessment arrangements.

PROCESS DEVIATIONS

All foreseeable process deviations should be considered for their effect upon the plant, particularly in relation to loss of containment. The most important process deviations are those concerning chemistry, pressure, temperature, flow rate and liquid level. According to the nature of the process under study, it may also be necessary to consider deviations of other variables, e.g. viscosity, surface tension and pH. Inspectors should scrutinise the foreseen process deviations in appropriate detail and make a technical appraisal of the occupier's competence to prevent loss of containment thereby.

Process Chemistry

Inspectors will need to ensure that the occupier has assessed correctly the sensitivity of the process to changes in process conditions. The effects on process chemistry of deviations of pressure, temperature, pH, viscosity, moisture content, contaminants, presence or absence of air/oxygen, etc. should be known, and safeguards taken, where necessary. The effects of omission or excess of each material involved in the process should have been adequately investigated, together with the consequences of contamination. For example, changes in pH may alter the reaction kinetics; excess catalyst may result in an uncontrollable exothermic reaction.

Possible consequences include unacceptable excursion of pressure, flow rate, filling ratio, need to vent or dump, foaming of vessel's contents, rollover, and irreversible metallurgical change to vessel.

Liquid Level

The limits of acceptable liquid level excursion in any vessel should be defined. Associated with liquid level are 2 fixed points: the maximum permissible load of vessels and the fill ratio. Advice for the differing fill ratios for various liquids is given in Industrial Codes and British Standard 5355. The liquid level will frequently need to be considered jointly with other factors.

Possible causes of an excursion in level include failure of overfill prevention device, poor communication systems between operators, e.g. ship/shore transfer without voice contact, opportunities for misrouting process streams, syphoning, loss of vacuum/pressure and surging. These might lead to spillage following overfilling or vessel rupture.

Flow Rate

An unacceptable flow rate may be one which is too high, too low or reversed in direction. Such excursions have been responsible for many serious incidents. Services are particularly susceptible to flow rate-related hazards.

Possible causes include furring up of pumps, pipelines and ancillaries; vapour lock; loss of pump prime; changes in differential pressure; and operator error. Possible consequences include temperature/pressure excursion (e.g. a fired process heater quickly ruptures if the process fluid flow is insufficient to maintain the correct temperature range of heat exchanger), and erosion or cavitation within pipes and pumps.

Criteria to be investigated include those for the use of flexible connections (and their renewal), gravity discharge and isolation devices, and those for the sizing and location of pipes, pumps, vents, flares, etc.

Impact

Consideration should be given to the effect of impact by foreign bodies. Possible causes include road and rail vehicles, seagoing vessels, missiles from other premises, objects dropped from cranes, mal-operation and poor maintenance of plant and proximity of hazardous plant. Particular attention should be paid to pipework connected to the base of any vessel which is not protected by a remotely operated isolation valve adjacent to the vessel.

Services

Consideration should be given to the effects of failure or other foreseeable excursion of services (steam, cooling water, instrument air, etc.) from specification. Excesses as well as reductions should be considered. Consideration should be given to foreseeable simultaneous excursions and to the need for duplication of facilities, particularly in cases where critical operations are involved.

Specialist inspectors may be asked to comment upon the safety integrity of 'independent' sources of electrical power.

Start-Up and Shut-Down

Routine start-up and shut-down constitute sufficiently large abnormalities, particularly of continuous processes, to warrant consideration as process deviations. Both these circumstances require extra vigilance on the part of operators and, in some instances, higher manning levels. The implications for start-up and shut-down of a change in reaction kinetics should be considered, especially where alternative reactions may supervene. Commissioning of new and rebuilt plants is generally the most difficult category of start-up.

OPERATOR ERRORS

Human error can contribute to plant failure and loss of containment at all stages — from plant design, through construction, commissioning, operation and maintenance, to eventual demolition. This section deals briefly with the aspects most relevant to plant operation and maintenance. Operator errors are essentially of 2 types: those of omission and those of commission. However, the underlying causes of the errors are usually more revealing.

Sensory Failure

If the operator does not receive correct or adequate information, he cannot be relied upon to act correctly. Such failures can result from instrument defects, poor gauge or panel layout, or lack of correct recognition of the available data due to inattention, lack of observation or inadequate training.

Failure in Logic

The operator, having correctly received the data, must then interpret it according to the prescribed logic. Failure in the logic which allows the operator to misinterpret the data is caused primarily by inadequate training. However, there are other factors which should not be overlooked, e.g. the stress created by unexpected or emergency situations, overloading due to a plethora of alarm signals or the effects of group decisions and pressures from colleagues.

Failure to Act Properly

Some human errors are of a random nature, when one simply forgets to perform some routine tasks (e.g. brushing one's teeth in the morning or checking the level on a storage tank). Training can do little to anticipate these errors, but good process design should minimise their effects and ensure that no single mistake should be capable of causing a catastrophe. Management systems, including effect supervision, checklists, permits-to-work, etc., can reduce the incidence of these errors.

EMERGENCY PROCEDURES

There must be an emergency shut-down procedure for all process plant, with appropriate means for stopping reaction, dumping, flaring, etc. Operators should have unambiguous guidance and adequate training on when to take such action, and on how and when to obtain assistance from senior staff responsible for the plant. The firms should have planned how storage vessels would be emptied, should the need arise.

Where spillage of a significant quantity of hazardous material can be foreseen, the firm should take measures to mitigate the consequences of such a spillage. These might include bunding, secondary (outer containment) tanks, catchment pits, dump tanks or control of spilled fluids by diversion walls and grading of the ground. The risk of accumulations of flammable liquid beneath process/storage vessels should be appraised. For some materials, an evaporation area at a suitable location is desirable. For others, particularly for toxic or cryogenic liquids, evaporation can be reduced by minimising the surface area of the bund or pit and covering it with a suitable blanketing medium or specific toxic remedy.

Ideally, there should be sufficient space between plant items to prevent unwanted interactions, for example, damage to adjacent plant as a result of an explosion. The integrity of control rooms for continuous processes should be ascertained. Within this constraint, the length and diameter of pipework should be minimised. Pipework should be routed so that it is not susceptible to damage, otherwise it should be protected. Vessel outlets and fixed pipework should have emergency isolation devices which can be operated from a safe place in the event of an incident.

Transfer operations to and from ship, rail and road tankers are particularly hazardous. They should be located so as not to endanger fixed plants and storages.

(Source Material: ILO Note on Techniques of Inspection of Major Hazard Works, ILO, Geneva.)

RELEASE OF FLAMMABLE AND TOXIC SUBSTANCES AND EFFECTS OF FIRES

by
Milos Nedved

DISPERSION IN ATMOSPHERE

Dispersion of hazardous and polluting materials in the atmosphere has been the subject of intensive interest for some decades and has resulted in the development of many different models. In the first instance, models of neutral (or tracer) plumes were of interest because these were relevant in examining the behaviour of pollutants discharged from vents and stacks. With the growth of interest in hazard analysis, the behaviour of clouds of material in which the vapour density is significantly different from that of the surrounding air became of interest. Usually, it is the denser-than-air behaviour which is of most interest because lighter-than-air gas will 'float' upwards and may, therefore, disperse harmlessly.

Dispersion of pollutants or small particles in the atmosphere is a function of the stability of air, wind speed and atmospheric turbulence. The standard categorisation system normally adopted for air stability is based on that derived by Pasquill. This scheme employs 6 or 7 categories to cover unstable, neutral and stable conditions: the latter are designated as A to F, or sometimes A-G. Neutral stability is designated category D and is the condition expected when there is total cloud cover, with or without precipitation. If the sun is shining, atmospheric turbulence is increased due to insolation or radiation. This gives rise to the less stable conditions classed as categories A to C inclusive. The most stable atmospheric conditions occur on clear, calm, cloudless nights (F, G)

Besides, the thermal effects on the atmosphere, turbulence is increased by higher wind speeds. These latter effects depend upon the surface roughness of the ground over which the atmosphere flows.

DENSE CLOUD DISPERSION
(Cox and Carpenter)

This model provides a reasonable estimate of the behaviour of those hazardous gases which may be significantly heavier than air, provided that the gas cloud is in a relatively flat environment, free from any significant obstructions.

OUTPUT: This method gives the spreading and downwind displacement of the cloud.

INPUTS: Initial cloud dimensions and concentrations, wind speed, atmospheric stability, material input rate or quantity, thermophysical data, characteristics of ground condition.

ACCURACY: Many comparisons and some tests have been conducted to validate this model. Usually, they can be calibrated to fit test data. This method is regarded as sufficiently accurate for modelling a gas cloud in a relatively 'flat' environment with no 'significant' obstruction. Care should be taken with situations where these conditions are not met.

APPLICATION: This method may be applied to examine the behaviour of hazardous gases or clouds which are heavier than air.

PASSIVE DISPERSION MODEL

METHOD:	Gaussian model for neutral clouds.
OUTPUTS:	The model gives cloud concentrations, both for instantaneous and continuous releases.
INPUTS:	(a) Released mass.
	(b) Dispersion parameters.
	(c) Wind speed.

Assumptions and Constraints:	Dispersion is based upon the assumption of Gaussian distributions of turbulence in the atmospheric boundary layer.
Accuracy:	Subject to a relevant selection of wind speed and stability category, short-term average concentrations of pollutants may be estimated within at least a factor of two of the experiment. This degree of precision is generally satisfactory for the purpose of a hazard assessment. However, due to limits in the applicability of dispersion coefficients, the method is generally applied to downwind distances in excess of 100 m. and less than 100 km. from the point of discharge.
Application:	The method determines the concentration distribution of clouds of neutral density.

BUOYANT SURFACE RELEASE MODEL

METHOD:	Briggs Plume Rise Model.
OUTPUT:	Release height estimate.

Method:
For a surface release of buoyant material, it is not necessarily clear that the material will lift off the surface under the action of buoyancy forces. The effects of turbulence, which may be intense near the ground due to friction effects and obstacles, may be dominant. For the cloud to lift off, the lower parts of the cloud edges must move inwards due to the external hydrostatic pressure acting against the spreading influence of the turbulent dispersion. Briggs (1976) has suggested that a criterion may be developed by comparing a characteristic lateral turbulent spreading velocity with a characteristic inward movement associated with buoyancy. This arises from considerations of the drawing in of the cloud sides near the surface as the bulk of the cloud starts to rise.

Assumptions and Constraints:
If the cloud does not lift off, it can only be treated as a passive tracer, using the appropriate dispersion models, e.g. the Gaussian model.

EFFECTS OF FIRES

The radiation effect of fires is normally limited to areas close to the source of the hydrocarbon (say within 200 m). In many cases, this means that the surrounding communities are not affected. However, there are some types of fires which would have a more pervasive effect.

Fires may be categorised as follows:

(a)   'Pool fire' (e.g. a tank fire or fire from a pool of fuel spread over the ground or water).

(b)   'Jet fire' from the ignition of jets of hydrocarbons or other flammable materials.

(c)   'BLEVE' (Boiling Liquid Expanding Vapour Explosion), resulting from the overheating of a pressurised vessel by a more minor primary fire which causes the vessel to explode and a large and very intense fireball to be produced.

(d)   'Flash fire', involving the ignition of a vapour cloud which does not explode, that is, the flame speed is not high, but the fire spreads quickly throughout the flammable zone of the cloud.

Fires affect the surrounding environment primarily through the radiated heat which is emitted. If the level of heat radiation is high enough, other objects which are flammable may themselves be ignited. Alternatively, living organisms may be burned by heat radiation, and thereby suffer either injury or death.

Damage associated with heat radiation may be assessed on the basis of the dose of radiation received. A measure of the dose received is the energy per unit area of the surface exposed to the radiation and the exposure duration. Alternatively, the likely effects of radiation may be estimated by using the power per unit area received. This latter approach is particularly relevant where the equilibrium between the power received and the power absorbed dictates the degree of damage that may be encountered.

FIRE DAMAGE

METHOD:	Fire damage estimates are based upon correlations with recorded incident radiation flux and damage levels.
OUTPUT:	Indication of damage as a function of incident radiation.
CONSTRAINTS:	Since damage estimates are based upon empirical evidence, the damage response characteristics should be updated as and when new evidence comes to light.

Method: Various tables have been created to set up criteria for damage to people and property from fire. Sometimes they are expressed in terms of radiation intensity, and sometimes as a power dosage. The effects on buildings, natural surroundings and equipment are measured in terms of the likelihood of ignition, particularly if wooden structures or buildings are in the vicinity. Spontaneous and flame-induced ignition values can be considered for various levels of radiation. The radiative or incident fluxes recorded are related to the level of damage and impact upon people, including plant personnel, based upon observations arising from actual incidents and large fires.

Outputs: The method provides estimates of fire damage, fatalities and injuries.

Inputs: Estimates of thermal flux at selected receptor points, using heat transfer phenomena.

Incident Flux (kW/m^2)	*Types of Damage Caused*
37.5	Sufficient to cause damage to process equipment. 100% lethality.
25	Minimum energy required to ignite wood at infinitely long exposures (non-piloted). 100% lethality.
12.5	Minimum energy required for piloted ignition of wood. Melting plastic tubing. 100% lethality.
4	Sufficient to cause pain to personnel if unable to reach cover within 20s: however, blistering of skin (first degree burns) is likely. 0% lethality.
1.6	Will cause no discomfort for long exposure.

Assumptions and Constraints: At the lower levels, where time is required to cause serious injury to people, there is often the possibility of escaping or taking shelter.

LITERATURE REFERENCES AND SOURCE MATERIAL:

1. Abraham, G. (1963). Jet Diffusion in Stagnant Ambient Fluid. Delft Hydraulics Laboratory, No. 29 Hydraulics Laboratory, Delft, Holland.

2. Cox., R.A. (1980). Methods for Predicting the Atmospheric Dispersion of Massive Releases of Flammable Vapour. Proc. Energy. Comb. Sci. Vol. 6, pp. 141-149, Pergamon Press.

3. Cox, R.A. and Roe, D.E. (1977). A Model of the Dispersion of Dense Vapour Clouds. Second International Loss Prevention Symposium, Heidelberg, 1977.

4. Cox, R.A. and Carpenter, R.J. (1979). Further Development of a Dense Vapour Cloud Dispersion Model for Hazard Analysis. Paper presented at the Symposium "Schwere Gase" at Battelle Institute, Frankfurt Am Main, September, 1979.

5. Cude, A.L. (1975). The Generation Spread and Decay of Flammable Vapour Clouds. ICHEME Course "Process Safety — Theory and Practice", Teeside Polytechnic, Middlesborough, 7-10 July, 1975.

6. Ellison, T.H. and Turner, J.S. (1960). Mixing of Dense Fluid in a Turbulent Pipe Flow. JFM. 8., 514-544.

7. Fauske, H.K. (1965). The Discharge of Saturated Water Through Pipes. CEP Symposium Series 61, p. 210.

8. Keffern, J.F. and Baines, W.D. (1963). The Round Turbulent Jet in a Cross-Wind. J. Fluid Mech. 15, 481-496.

9. Lapple, C.E. (1943). Isothermal and Adiabatic Flow of Compressible Fluids. Trans. Am. Inst. Chem. Engrs. 39, 385.

10. Moody, L.F. (1944). Trans. Am. Inst. Soc. Mech. Engrs. 66,671.

11. Morrow, T.B. et al. (1980). The Dispersion of Chemical Vapours Emitted from Marine Chemical Carriers. Third International Symposium on Loss Prevention and Safety Promotion in the Process Industries, Basle.

12. Ooms, G. (1972). A New Method for the Calculation of the Plume Path of Gases Emitted by a Stack. Atmospheric Environment, Pergamon Press, 1972.

13. Ooms. G., Mahieu, A.P. and Zelis, F. (1974). The Plume Path of Gases Heavier Than Air. First International Loss Prevention Symposium, The Hague/Delft, 1974.

14. Pasquill, F. (1974). Atmospheric Diffusion. 2nd Edition, Ellis Horwood, Chichester.

15. Perry, J.H. (1963). (Ed). The Chemical Engineers Handbook. 4th Edition, McGraw-Hill.

16. Perry, R.H. and Chilton, C.H. (1973). Chemical Engineers Handbook. 5th Edition. McGraw-Hill. Kogakusha.

17. Peterson, R.L. and Cermak, J.E. (1979). Plume Rise for Varying Ambient Turbulence, Thermal Stratification and Stack Exit Conditions in Numerical and Laboratory Evaluation. Fourth International Conference on Wind Engineering, Fort Collins, Colorado.

18. Turner, D.B. (1969). Workbook of Atmospheric Dispersion Estimates. U.S. Department of Health, Education and Welfare. PHS Pub. No. 995-AP-26.

19. Briscoe, F. and Shaw, P. Evaporation from Spills of Hazardous Liquids on Land and Water. SRD R100 1978.

20. Cox, R.A. and Roe, D.R. A Model of the Dispersion of Dense Vapour Clouds. Loss Prevention and Safety Promotion, p. 359 (1977).

21. Cox, R.A. and Carpenter, R.J. Further Development of a Dense Vapour Cloud Dispersion Model for Hazard Analysis. Heavy Gas and Risk Assessment, ed. Hartwig, S., (1980).

22. Hall, D.J., Barrett, C.F. and Ralph, M.O. Experiments on a Model of an Escape of Heavy Gas. Report LR 217, Warren Spring, Stevenage, Herts. (1975).

23. Picknett, R.G. Field Experiments on the Behaviour of Dense Clouds. Report Ptn. 1154/78/1, Chemical Defence Establishment, Porton Down, Wilts.

24. A.P. van Ulden. On the Spreading of a Heavy Gas Released Near the Ground. First International Loss Prevention Symposium, The Hague/Delft (1974).

25. Flare Radiation. API RP 521.

26. Hajek, J. and Ludwig, E. Petrochem. Eng. Part 1. (1960).

27. Jost, W. Explosion and Combustion Processes in Gases. McGraw-Hill, New York (1946).

28. Zabetakis, M. and Burgess, E. Bureau of Mines Report. Invest 5707 (1961).

29. The World Bank. Manual of Industrial Hazard Assessment Techniques. The World Bank, Washington D.C. (1985).

EXPLOSIONS

by
Milos Nedved

The definitions of "explosion" are:

"A rapid increase of pressure in a confined space, generally caused by the occurrence of exothermic chemical reactions in which gases are produced in large amounts".

"A rapid release of energy".

The source of energy is unimportant.

EXAMPLES:

(a)   heating a tin of beans;

(b)   stored energy in a compressed gas;

(c)   nuclear device;

(d)   hydrocarbon oxidation;

(e)   thermal expansion; and

(f)   polymerisation.

All these can give rise to explosions, but the energy release must be sudden, giving rise to a high concentration of available energy. This concentration is then dissipated, the energy being used in various forms of external work:

(a)   fragment effect; and

(b)   shock wave generation, etc.

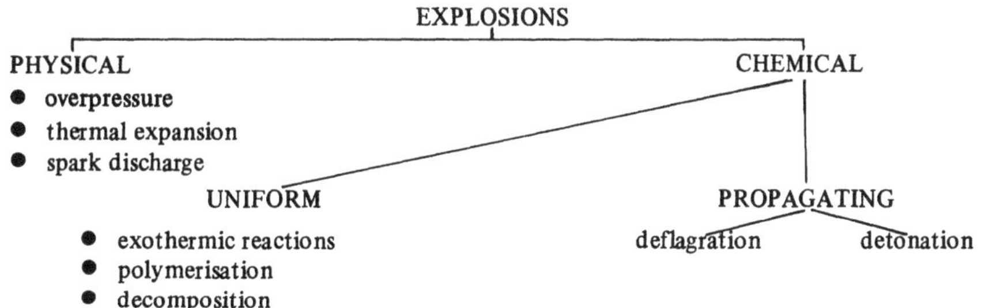

ASSESSING THE POSSIBLE EFFECTS OF AN EXPLOSION

Estimating Energy of Explosion

(a)   Based on Isothermal Expansion

$$E = 1.26 \; V \; [P_1/P_0] \; [T_0/T_1] \; RT_1 \; \ln \; [P_1/P_2]$$

V —   volume of a vessel in ft^3.

P_1 —   pressure of compressed gas in atmosphere.

P_2 —   final pressure of compressed gas in atmosphere.

P_0 —   standard conditions: 1 atmosphere.

T_1 —   temperature of compressed gas in $^\circ K$.

T_0 —   standard temperature: $273^\circ K$.

R —   gas const.: 1.987 cal/gm mole $^\circ K$.

1.26   conversion factor, ft^3 to gm moles.

E —   energy in cal.

Example: What is the energy stored in 1 ft^3 of gas at 2000 psi and $20^\circ \; C^2$

$$E = 1.26 \; x \; 1 \; x \; \frac{2000}{14.7} \; x \; \frac{273}{293} \; x \; 1.987 \; x \; 293 \; x \; \ln \frac{2000}{14.7} = 4.6 \; x \; 10^5 \; cal$$

The standard for energy release for TNT is 5×10^5 cal/1b.

Therefore, the energy stored in 1 ft^3 gas at 2,000 psi, and $20^\circ C$ is equivalent to the energy of 0.92 1b. of TNT.

(b)   Based on Chemical Potential (Method for C, H, N, O Materials)

1.   Write reaction products. Assign available oxygen to form CO, then H_2O, then CO_2, in that order.

2.   Assign standard heats of formation to products and reactants.

3.   Calculate heat of reaction.

4.   Compare the heat of reaction for TNT.

Example: Estimate the heat of reaction for DNT at $25^\circ C$ and 1 atm. Compare with similar calculations for TNT.

The applicable heats of formation are:

TNT: 8.6   Kcal/mole
DNT: 3   Kcal/mole
CO: 26   Kcal/mole

For DNT

$$C_7 H_6 N_2 O_4 \rightarrow 4CO + 3C + 3H_2 + N_2$$

$$\triangle Hr = 4 \ H_f \ (CO) - H_f \ (DNT)$$

$$\triangle Hr = 4 \ (-26) - (-3) = - \ 101 \ Kcal/mole$$

M.wt. of DNT = 182

$$Q = 555 \ cal/g$$

For TNT

$$C_7 H_5 N_3 O_6 \rightarrow 6 \ CO + C + 2.5 \ H_2 + 1.5 \ N_2$$

$$\triangle Hr = -147. \ 4 \ Kcal/mole$$

M.wt. of TNT = 227
$$Q = 650 \ cal/g$$

TNT equivalent of DNT $\dfrac{555}{650}$ x $100 = 85\%$

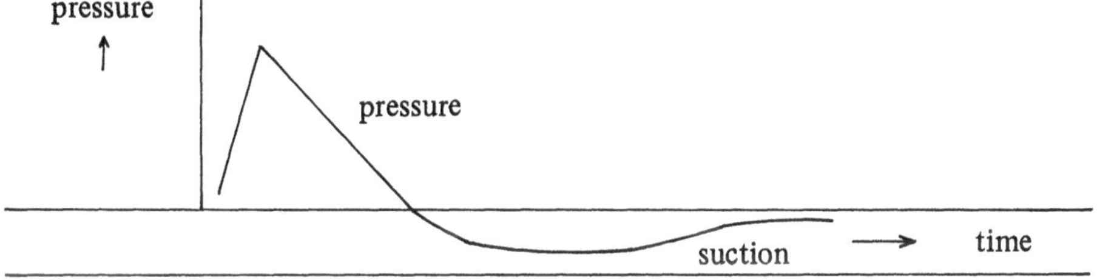

CHARACTERISTICS OF AIR BLASTS FROM DETONATING: 1kg of TNT

Distance (m)	Shock Velocity (m/s)	Peak Static (atm)	Peak Negative (atm)	Reflected (atm)
0.2	3,690	159	1.0	1,460
1.0	1,060	11	0.7	63
5.0	370	0.28	0.065	0.64
10.0	348	0.114	0.033	0.24
100.0	332	0.007	0.004	0.014

EXTIMATING BLAST EFFECTS

1.   Establish characteristics of shock wave for standard explosive.

   (a)   Static pressure Pso (so = side on) (can be measured in a direction normal to the propagation direction).

   (b)   Negative pressure – P.

   (c)   Dynamic pressure $q = 0.5 \, \varsigma \, v^2$

      ς – density of gas

      v – velocity

      Analogous to wind meeting an object.

   (d)   Total or reflected pressure.

      $Pr = 2 \, (P + q)$

See: Chapter 3 – Glasstone S.: The Effects of Nuclear Weapons. U.S. Government Printing Office, 1962.

2.   Scaling Law

   A given pressure will occur at a distance from an explosion that is proportional to the cube root of the energy yield.

$$D = Do \left(\frac{W}{Wo} \right)^{.333}$$

3.   Using TNN equivalent of material in question (discussed previously) based on:

   (a)   energy of confined gas for vessel rupture;

   (b)   calculation of chemical potential for detonating materials; and

   (c)   experimental determination.

4.   Relate blast effects to structural strength of material and physiological damage.

FAILURE OF STRUCTURAL ELEMENTS DUE TO BLAST PRESSURE

Structural Element	*Peak Overpressure (psi)*
Glass windows	0.5 – 1
Corrugated panelling	1 – 2
Cinder block walls	2 – 3
Wooden telephone pole	5

PHYSIOLOGICAL EFFECTS OF BLAST PRESSURES

Physiological Effect	Peak Overpressure (psi)
Knock personnel down	1
Eardrum rupture	5
Lung damage	15
Threshold fatalities	35
50% fatalities	50
99% fatalities	65

UNCONFIRMED VAPOUR CLOUD EXPLOSION (UVCE)

The published data suggests that only a fraction of the flammable cloud takes part in the explosion — around 10 per cent (or less).

Pressures of between 20-30 psi are possible.

Pressures of 10-15 psi will severely damage, if not demolish, most buildings.

HEALTH HAZARDS FROM CHEMICALS IN THE WORKPLACE
by
Peter Hollingworth

INTRODUCTION

This section gives some idea of the potential effects on humans of some chemicals commonly encountered in the workplace. It is not meant to be encyclopaedic, rather to stimulate Factory Inspectors to think of the effects of chemicals on humans.

It is the duty of all who work with chemicals to know exactly what chemicals are being handled, what their effects are and what steps should be taken if accidents occur during storage, transport or use of the chemicals.

The subject will be looked at under several headings:

A. Checklist of chemical hazards to health.

B. Host factors modifying an individual's response to a particular chemical.

C. Special preventive measures in industry.

D. Sensory perception and chemicals.

E. Some chemical hazards associated with the textile industry.

F. Check list of physical hazards to health.

G. A detailed look at common chemicals — Solvents.

H.   — Fluorides,

I.   — Acidic Gases and Phosgene.

J.   — Ozone.

K.   — Styrene and Polystyrene.

L.   — Mercury.

M.   — Chromium.

N.   — Cyanide.

O.   — Vanadium.

A. CHECKLIST OF CHEMICAL HAZARDS TO HEALTH

Physical Modes:

Gas.
Liquid.
Vapour.
Dust.

Entry Modes:

Inhalation
Ingestion.
Percutaneous.

Toxicity:

Acute.
Chronic.
Cumulative.
Teratogenic, Embryotoxic.
Carcinogenic.

Action:

Direct.
Indirect.
Facilitate other compounds' action.

Site of Action:

Local.
Systemic.
Metabolic pathway.

Breakdown and Excretion:

— Metabolites also toxic.

— Compromised metabolic pathways.

— Competitive inhibition.

— Modified by host factors.

B. HOST FACTORS MODIFYING AN INDIVIDUAL'S RESPONSE TO A PARTICULAR CHEMICAL

The effects of a particular chemical are not always identical on various individuals. A number of factors in the host may modify the way in which the indivi-

92

dual responds to exposure to a chemical. There are four main groups of such "host factors", viz:

(a)   Genetic:

- Ethnic – such variations as liver enzymes, haemoglobins.
- Cilia clearance – the "cleaning" action of such cells is varied.
- Macrophage function – the cells responsible for some of the engulfing and neutralisation in the lungs of various foreign bodies and particles.

(b)   Acquired:

- Cilia clearance damaged, e.g. by cigarettes, temperature.
- Liver function altered, e.g. by alcohol, medication.
- Renal function altered, e.g. by drugs, medication.
- Respiratory rate altered, e.g. by drugs, temperature – these are usually only temporary factors.

(c)   Anatomical:

- Especially as regards nasal passages and airways.

(d)   Immunological:

- Increased sensitivity, e.g. asthma, where exposure increases the host reaction to the chemical.
- Decreased sensitivity, e.g. byssinosis, where exposure to the noxious agent decreases the host reaction.

Treatment

When considering the ways in which chemical exposure of the individual may be managed, there are three main types, viz:

(a)   specific antidote.

(b)   specific support.

(c)   general.

Prevention

Much of the industrial exposure of individuals to chemicals could be prevented by careful usage. Areas needing special attention are:

(a)   Storage and handling – a very common area for contamination.

(b)   Personal protection equipment.

(c)   Safe use — Spray drift and wind must be considered.

> Spray and eating areas — the latter must be free of chemical contamination.

> Contact with residues of chemicals, e.g. chlorinated insecticides, which may persist on surfaces with which the worker comes into contact.

C.   SPECIAL PREVENTIVE MEASURES IN INDUSTRY

Enclose the operations, isolating the operator from the chemical.

Exhaust safely and flush with fresh air before the operator enters impervious surfaces, e.g. preventing mercury being absorbed into wooden bench tops.

Adequate and flushable drains with competent water traps.

Vacuum toxic dusts or damp sweep/mop rather than disturb dust.

Bury in sealed containers, e.g. disposing asbestos safely.

Decreased working temperature of the process may decrease the volatility of some chemicals.

Wherever possible, replace toxic with less toxic substances.

When all else fails, observe Threshold Limit Value (TLV), remembering *NOT* all operators will be unaffected even at the TLV level.

(Remember effects of overtime or a long work-span with break only after a long interval).

(Remember most TLV are based on an 8-hour day, five-day week, with a two-day break before recommencing.)

Education

Education is necessary for operators and managers, and must consider both normal situations and possibly abnormal chemical reactions. Special attention for education in:

(a)   handling/storage.

(b)   normal use.

(c)   dangerous incidents or abnormal reaction.

Surveillance

In order to ensure that despite the foregoing the workers and surrounding population are in fact safe, there is need for on-going surveillance and to monitor — environmental; biological — may be the workers, may be the surrounding population.

94

Poisoning

Poisoning or chemical exposure at work can present particular and special problems in pregnancy, both for the mother and foetus.

(Remember problems of mixed industrial exposure and possible therapeutic or social medication, where the chemical effects on the body may be additive, antagonistic or occasionally synergistic.)

D.   SENSORY PERCEPTION AND CHEMICALS

When considering safety at work, it is important that an employee's perception of his surroundings is not in any way distorted. Distortion may take the form of misperception by the sensory organ, such as eye or ear, or it may take the form of interference with brain functions, thus permitting misinterpretation of the message sent by the sensory organ. Below are some examples of agents which may be encountered in the workplace which have the potential to either affect the sense organ itself or affect brain functions. Most of them are chemicals, and where an element is nominated not all compounds will be involved: in some cases only toxic doses of the element or its salts produce effects; in some cases the effects only occur with chronic exposure, in others with acute exposure; in many cases the effects are reversible. The list is not exhaustive but will give some idea of the range of potential effects on the operator.

Vision

Blindness	Methyl Alcohol (decreased fields; dilated pupils; photophobia; decreased acuity)
	Ethyl Alcohol.
	CO. Hg. No.
	Flying — Centrifugal Forces.
Chromatopsia	Flying — Red Out.
	Snow Blindness — After Image — Red — also occasionally seen in users of bright screens or monitors.
	Conjunctivitis or Oedema — Rainbow.
	Yellow — Auramine, DDT, Picric Acid, Nitrites
	Blue and Yellow — Amyl Alcohol, Trichlorethylene
	Grey — Blue — Amines
	Variable — Bromides, Co, Pb.

Diplopoe	Narcosis.
Nystagmus	Radar.
	Chlorinated Hydrocarbons, e.g. Trilene.
	Ethylene Glycol.
	Poor light, e.g. miners.
	Pb. Li. Mn.
Optic Atrophy	Leptospirosis
	As. Pb. P. Thallium. Tin alkyl cps.
	CCl_4. CS_2. Trilene.
Photophobia	Ultrasound
Pupils	Dilated in cornfied workers (Jimson Weed contains Stramonium).
	Narcosis.
Scotomata	Decompression sickness.
	CS_2. Pb. Hg. Thallium.
	CH_3OH. CH_3Br.
Visual Disturbance	Altitude.
	Vibration 10-25 HZ.
	Acceleration – Red/Grey/Black at Blast.
	As. Pb. Li. CO_2 CS. CCl_4.
	–ve Decompression.
	Dieldrin/Organophosphates/Phosphorus.
	Methyl – Acetate/Bromide/Chloride.
	Gasolene. Nitrobenzene.
Field Constriction	As. Hg. CS_2. CCl_4. CH_3OH.
	O_2 poisoning.
Haloes	Chlorine Dioxides, Ethylene Diamine, etc.
Hallucinations	Bromides.
	CO. Pb. C_2H_5 OH.

Mental

Confusion

-     Due to Anoxia – Anaemia
                         Altitude
                         Heart disease
                         Methaemoglobin

- Deceleration.
- Liver damage – poisoning.
- Oxygen posioning.
- Camphor. Mn. NO_2. Organophosphates.
- Salicylic Acid.
- Parathion – apple pickers.

Convulsions

Lots and lots of chemicals. Important ones are:

- Organophosphates.
- Solvents. CCl_4. Trilene. Perchlorethylene.
- CO & O_2 poisoning.
- Alcohol.
- Salicylic Acid.

Dizziness and Drowsiness

Lots and lots of chemicals. Important ones are:

- Alcohols.
- Solvents.
- Organophosphates.
- Petrol/Ether/Benzene/Nitrobenzene.
- Xylene/Toluene.
- Acetone/Acetates.
- Lead.
- Salicylic Acid.
- Anoxia.

Encephalopathy

- Lead.
- As. Mn. Au.
- Cyanides.
- Ethylene Glycol. Toluene.

—   CO. H$_2$S.

—   Compounds causing Methaemoglobin.

Insomnia

—   Stress.
—   Circadian rhythm disturbances, especially shift change and travel.
—   Brucellosis and Tetanus.

Irritability

—   Hypoxia — altitude.
—   Hypothermia.
—   Ultrasound. Noise.

Impaired Memory

—   Hypoxia, Decompression.
—   Br. Hg. CO. Organophosphates.

Headache

—   Several hundred chemicals or industrial conditions can cause this very protean symptom.

Personality Changes

—   Altitude — Hypoxia.
—   As. Pb. Mn. Hq.
—   CO. CS$_2$. Camphor.
—   Chlorinated Pesticides.
—   Kerosene, Napthalene, Pyrethrins.
—   Sensory deprivation, e.g. White Out Syndrome

Musculoskeletal

Facial Weakness

—   Some pesticides.
—   Benzene Pb. Mn. Hg. Cd. CO. Trilene.

Neuropathy

—   Longer Neurones are affected first and Reflex Arcs are interrupted with loss of reflexes.

—   Decompression. Vibration. Cold.

— As. Pb. Hg. P. Thallium.

— Some Halogenated Hydrocarbons.

Paralysis

— Decompression.

— Ultra Sound.

— Leptospirosis.

— Envenomation.

— Ba. Pb. Mn. Hg. Thallium.

Parkinsonism

— Pb. Hg. Mn. CS_2.

Parasthesiae

— As. Pb. Thallium.

— Hypoxia. Cold. Vibration.

— Aniline. HF. H_2S.

— Ch_3Br Ch_3Cl. Toluene, Xylene.

— Vinyl Chloride.

— Venoms.

General Weakness

— Lots and lots of chemicals and physical conditions, e.g. heat strain, heavy metals.

Wrist Weakness, Palsy

— Heavy metals.

Tremors

— Heavy metals. Pesticides.

— Salicylic Acid.

— Vanadium.

Festination

— Mn.

E. SOME CHEMICAL HAZARDS ASSOCIATED WITH TEXTILE INDUSTRY

It is frequently forgotten that the textile industry uses a very wide range of chemicals. All too frequently, it is only the physical hazards that are remembered: thus all students know of the risks of byssinosis in cotton workers, but ignore the associated chemical risks once the cotton cloth has been manufactured.

A general idea of the dyeing process is outlined, but there are variations in this basic theme.

Dyeing

Cotton: preparation for dyeing.

(a)   Shearing machine to trim loosely adhering fibres.

(b)   Pass over gas flame.

(c)   Desizing — pass through Diastase solution.

(d)   Scour in a Kier with dilute NaOH, $Na_2 Co_3$ or Turkey Red Oil Leave 8-12 h. @ high temperature and pressure.

Or, if coloured woven material:

(a)   Removal by hypochlorite solution.

(b)   Aired.

(c)   Washed.

(d)   Dechlorinated with Sodium Bisulphite.

(e)   Washed.

(f)   Scoured with dilute HCl or dilute $H_2 SO_4$.

Wash and ready for dyeing.

Wool: preparation for dyeing.

(a)   Scour — Soap and Soda Ash.

(b)   Bleach with $H_2 O_2$ or SO_2.

(c)   Neutralise acid gas by passing fabric through $Na_2 CO_3$.

Some dyes are an organic base. Solvents are used to make an emulsion with them.

Storage of Solvents — fire resistant sheds, bunds, ventilation.

Bleach with Hypochlorite solution., or gaseous Chlorine, or Bleaching Powder — all potentially hazardous.

CS_2 used as Solvent in Viscose process — needs care in use.

Cleaning of machine by hazardous chemicals may occur.

Hazards of working in a Kier — scalding when the discharge from another part enters the Kier.

Dyeing Industry

General chemical hazards include:

Acids.
Alkalis.
Bleaches.
Detergents.
Moisture.
Solvents.
Zinc Chloride.

Dyes and Pigments

The following substances or their salts are commonly encountered:

Auramine.
Benzidine base pigments ⎫
Fuchsin. ⎬ All cause bladder cancer.
Magenta. ⎭
Antimony causes skin irritation.
Arsenic causes skin darkening, eczema, nasal perforation.
Barium causes eczema.
Bromide causes browning of the skin, rashes and eruptions.
Carbon black causes cancer of skin.
Cadmium causes chemical pneumonitis.
Chromium causes skin ulcers, dermatitis, nasal perforation.
Lead causes neurologic and haematologic changes.
Mercury causes skin irritation, neurologic changes.
Nickel causes dermatitis.
Selenium causes dermatitis, neurologic changes.
Sodium silicate causes skin thickening and ulcers.
Trisodium phosphate causes skin blisters and ulcers.
Zinc causes skin ulcers (Zn chromate is a carcinogen).

Fur Industry

(a)   Hide is coarse fleshed.

(b)   Soften hide by soaking in brine.

(c)   Spin dry.

(d)   Fleshing: mechanically done — risks of trauma to operators.

(e)   Soak in Alum solution — either with HCl or H_2SO_4.

(f)   Dry.

(g)   Oil.

(h)   Clean off excess oil and moisture by revolving in a drum with sawdust.

If Dyeing needed:

(a)   Soak in weak Alkaline ($NaHCO_3$).

(b)   Soak in Mordant, e.g. Ferric Sulphate.

(c)   Steep in Dye, e.g. Aniline for shoes.

(d)   Rinse/dry/sawdust as in (h) above.

Hazards include:

> Acids.
> Alkalis.
> Alum.
> Bacteria, especially Anthrax.
> Bleaches.
> Chromates.
> Dyes.
> Formaldehyde.
> Fungi.
> Lime.
> Oils.
> Salt.
> Solvents.

Felting

Felt usually made from jute or shoddy.

Risks — fire, dust, anthrax.

Hat-making

"Carrotting" — originally $Hg(NO_3)_2$.

Now   (a)   Oxidising agents — H_2O_2, Perchlorates and Organic Acids.

(b)   Oxidising agents — Na or Ba Peroxides + H_2SO_4.

(c)   Reducing agents — Salts of Phosphorus.

(d)   Arsenic and its compounds.

Waterproofing

    Alum.
    Formaldehyde/Melamine Resins.
    Oils.
    Paraffins.
    Pitch.
    Rubber.
    Solvents.
    Waxes.

Shrinkproofing

Use *Ficin* for shrinkproofing wool or removing Gelatin from sized thread.

Ficin is dangerous because of its strong proteolytic action, hence it causes irritation of skin, eyes, mucous membranes. If taken orally, it causes purging.

Castor bean dust — causes contact dermatitis.

Dry cleaning

Some of the hazards:

    Acetic Acid.
    Ammonia.
    Amyl Acetate.
    Benzene.
    Carbon Tetrachloride.
    Chlorobenzene.
    Dichloropropane.
    Diethylether.
    Dusts.
    Methyl Alcohol.
    Nitrobenzene.
    Perchlorethylene.
    Sizing Compounds.
    Trichlorethylene.
    Turpentine.
    Waterproofing Compounds.
    White Spirits (Stoddard Solvent).

F.   CHECKLIST OF PHYSICAL HAZARDS TO HEALTH

The physical health hazards which may affect safety by altering body perception are listed below (There are of course many physical hazards which do not directly affect perception):

     Vibration.
     Noise.
     Ultrasound.
     Illumination.
     Ultraviolet radiation.
     Infrared radiation.
     Ergonomic factors.
     Heat.
     Cold.

Below are examples of the types of symptoms which may be caused by physical factors in the workplace. Some will only occur at extreme levels, others are commonly encountered in industry:

Ultrasound

     Deafness.
     Headaches.
     Raised blood pressure.
     Fever.
     Eosinophylia.
     Labyrinthine disturbances.
     Photophobia.
     Fatigue.
     Paralysis.

Noise

     Deafness.
     Raised blood pressure.
     Cardiospasm.
     Raised catecholamines.
     Muscle tension.
     Headache.
     Loss of concentration.
     Altered field of vision.
     Altered colour perception.

G.   A DETAILED LOOK AT COMMON CHEMICALS

SOLVENTS

Solvents are freely available both in industry and domestically, and are increasingly abused, especially by young people. Some of these abuses lead to death.

Abuses

Solvent-associated deaths in U.K. in 1985:

Butane	28%
Adhesives	23%
Aerosols	15%
Fire extinguishers (esp. B.C.F.)	5%
Others	26%

Frequently-abused solvents include:

B.C.F. = Bromochlorodifluorometane.
Vinyl Chloride.
Dimethyl Ether.
Propane.
Isobutane.
Heptane.
Benzene.
Xylene.
Toluene.
Petrol.
Petroleum Ether.

Uses

Widespread throughout industry and home. Industrial uses include:

Glue.
Fibreglass.
Electronics.
Refrigeration.
Dry cleaning.
Paint manufacture.
Cleaning/degreasing, especially electrical.
Paint strippers.
Propellants.

Usually used as mixtures — cost, safety factors dictate the mix.

A common example of a mixed solvent for use in a degreasing bath would be:

Methylene Chloride (Hi cost, Hi solubility, Mod. Tox, V. Lo Flamm).
Perchlorethylene.
Stoddard Solvent (Lo Cost, Lo Solubility, Mod. Tox, Flamm).

Stoddard Solvent = White spirits

Flashpoint 100-110°F.

Ignites when the other chlorinated HCs have evaporated.

Inhalation affects the central nervous system and can lead to coma.

Inhalation or ingestion causes burning sensation, and later pneumonitis and sterile bronchopneumonia.

Glycol Ethers

Used in paints, coatings, inks and surface cleaning.

10 minute exposure not to exceed 3 x TLV.

But absorbed through intact skin — there is no TLV for skin absorption.

Brain Damage

Solvents cause brain damage.

Early retirement on psychiatric grounds is twice as likely in workers who have had occupational exposure to solvents. These early retirees include pre-senile dementia sufferers.

Two classes of brain damage are reported commonly;

(a)   organic affective disorder;

(b)   chronic toxic encephalopathy.

Xylene-based glue can cause status epilepticus, a medical emergency.

Reference: W.H.O. Environmental Health No. 6 — 1985.

Other Effects of Solvents on the Body

Pregnant women exposed to Chlorinated Hydrocarbons have a greater chance of bearing children who develop leukaemia in childhood. The exposure of the father is irrelevant.

Renal damage, especially with Carbon Tetrachloride.

Hepatic damage is associated with many solvents, and may take a year before recovery is complete. Occasionally, hepatic damage leads to complete hepatic failure and death.

Myocardial Sensitisers

It is becoming increasingly recognised that many solvents have the ability to sensitise the heart muscle to Catecholamines.

106

Commonly implicated: Trichloroethylene.
Trichlorotrifluoroethane.
Trichlorofluoromethane.
Halothane.

Solvents sensitising the Myocardium are mainly Halogenated Hydrocarbons.

Myocardium is sensitised to endogenous and exogenous Catecholamines, i.e. both to those produced by the body as well as those administered (e.g. by a doctor or nurse).

Biological Monitoring of solvents is frequently possible.

Examples:

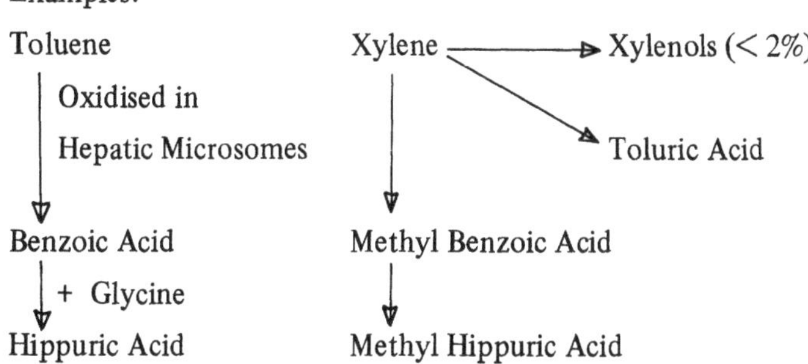

(Note that in many countries Benzoic Acid is a degradation product of Sodium Benzoate used in food preservation. It is, thus, not possible to *quantitatively* estimate Toluene exposure by only measuring urine output of Hippuric Acid.)

Perchlorethylene

Uses — very many. Include: Common solvent in degreasing baths, etc.
Cleaning surfaces for adhesives or belting.

TLV is 50 p.p.m.   Toxic ≃ 230 p.p.m.

(Fatal ≃ 0.4 mg/L   Cardiac 0.04 mg/L)

Action — Liver and kidney
— Demyelination — cardiac arrhythmias

Acute Effects		*Chronic Effects*
Ingestion ⎤	Dizziness, ataxia	Neuropathy, especially peripheral
Inhalation ⎦	Headache	Later → Brain
	"Drunkenness"	

Maternal milk → Babe → Jaundice in the *baby* from *maternal exposure* to Perchlorethylene.

Ventricular arrhythmia may occur if in chronic exposure myocardial sensitisation has occurred.

Exercise is very dangerous as Catecholamine is released by the body and can affect the sensitised heart muscle.

Do *NOT* give Adrenaline or other Catecholamines, as they can prove fatal.

Myocardial sensitisation is a two-stage process: (a) sensitisation of heart muscle, and (b) Catecholamine release.

Treatment: O_2

May have renal and/or hepatic shut-down, which may need special treatment.

Recovery/Survival

Liver may take 12/12 to recover.

If not early death ($< 24/24$), usually survive.

Diagnosis

Most retrieved in expired air — not easily tested in the workplace.

Urine metabolites, e.g. Trichloracetic Acid 3 per cent excreted is specific and readily detected.

Found in blood *BUT* takes most of week to stabilise ∴ measure towards end-of-shift at end-of-week — technically, a harder test.

Prevention

Exclude those with known liver or renal impairment.

Regular monitoring R.F.T. and L.F.T.

Personal protective gear.

FLUORIDES

Examples of industrial uses — cleaning metals.
- glass etching.
- extraction of ore.
- fluoridation of water (toothpaste).
- superphosphate production (fluorapatite).
- rodenticide.
- fumigant (methyl sulforyl fluoride).
- aluminium reduction (sod. al. fluoride).
- alkylation high octane petrol (aviation).

Fatal dose Na F 5-10 mg/Kg of F.

Fatal plasma level $\simeq$ 3 mg/litre

In workers   10 mg/day Na F $\rightarrow$ Osteosclerosis, i.e. very hard, abnormal bones.

In Osteoporosis   60 mg/day Na F tolerated.

F. interferes with Ca metabolism, enzyme mechanisms, lowers body Mg levels.

Acute Toxicity

(a)   *H.F.*   Inhalation in 1-2 hours, irritation of respiratory tract.
                in 1-2 days, Pulmonary Oedema. May be 30 days before clear.

      Ingestion — Early irritation of the gastro-intestinal tract.
                — Vomiting, abdominal pain, salivation, diarrhoea.
                — C.N.S. effect can lead to respiratory paralysis.

      Skin and
      Eye   — Very painful burns: penetrate if treatment inadequate: Eschar forms with corrosion continuing under ($\simeq$ Chrome ulcers).

(b)   Fluorides — Salts

      Inhalation causes Narcosis.
                       Methaemaglobinaemia.
                       Kidney damage.
                       Liver damage.

Chronic Toxicity

H.F. Inhalation — Asthmatic type problems.
                 — Altered Immunoglobulins found in some potrooms.

Workers in aluminium refineries.

Ingestion — Anaemia.
          — Weight loss.
          — Brittle bones.
          — Tooth and bone mottling (including babes and kids < 9 years)
          — Fluorosis = Osteosclerosis + Osteophytes, with roughening and chalky patches on the cortex of the bones.

Treatment

Skin — Copious lavage with water.
     — Na Ha CO$_3$. 1:2,000 Iced Zephiran.
     — Calcium Gluconate — cream or injection to neutralise the Fluoride.
     — Mg Oxide (used in U.S. to similarly precipitate fluoride).

Ingestion   — Drink milk – calcium content is high.
            — Drink Ca Gluconate or Mg SO_4 to form neutral flourides.
            — If low serum Mg R_x I.V. Magnesium – but only under laboratory monitoring.
            — May need I.V. Calcium Gluconate.

Inhalation — Supportive, O_2, may be I.V. Mg. or Ca. Gluconate.

Outcome:

   Acute

Inhalation: If survive 3-4 days, usually O.K.

Ingestion: If survive 48 hours, usually O.K.

Burns: May take months to heal ± surgery, skin grafting.

   Chronic

Fluorosis: Often recovery period is over a year.

Allergic Asthma: Never recover.
                Retain sensitivity for life.
                Cross-sensitivity can occur.

Organic Fluoride Compounds

Fluoracetic Acid (FCH$_2$ COOH) v.v. toxic

1 drop on skin can kill. Use as rodenticide.

Poisoning commences with apprehension – Grand Mal within 2 hours.

May lead to unconsciousness and fatal ventricular fibrillation.

Fluoroacetate → Fluorocitric Acid, Inhibits Aconitase ∴ stops Tri Carboxylic Acid cycle.

Fluorocarbons

Fluorocarbons used as refrigerants. Safe until heat destroyed.

Fluorochloroethanes.

Fluorochloroethylenes.

Commonest used commercially as refrigerants are:

    Dichloro Difluoromethane = Freon 12 (also a Propellant).

    Trichloro Fluoromethane = Freon 11 (also a Propellant).

    Dichlorotetrafluoroethane = Freon 114 (also a Propellant).

Volatile Fluorocarbons have similar effect to volatile Chlorinated Hydrocarbons, but they are less marked.

Hence: Narcosis, headache, dizziness.
Surgical anaesthesia.
Skin defatting.
Cardiac arrhythmias.
Myocardial sensitisation to Catecholamines.

They have variable mutagenicity and teratogenicity (e.g. $CHClF_2$ + ve).

The potential for mutagenicity or teratogenicity probably applies to incompletely Halogenated Hydrocarbons.

Note: (a) Lower Lipoid solubility makes fluoridated HC less toxic than Chlorinated HC.

(b) CF bond more stable than CCl bond.

(c) Fluoroalkenes are very variable toxicity.

Perfluoroisobutylene $(CF_3)_2$ and C: CF_2 are more toxic than Phosgene.

c.f. Vinyl Fluoride and Vinylidene Fluoride v. low toxicity.

(d) Fluoroalkenes, e.g. Tetrafluoroethylene, can produce liver and kidney damage.

c.f. Fluoromethane and Fluoroethanes which do not.

Polymeric Fluorocarbons

Teflon is a Polymeric Fluorocarbon — Polytetrafluorethylene.

V. low coefficient of friction, hence use in bearing.

No solvents attack it and only attacked by F and molten Alkali metals.

Good electrical properties.

$> 400°C$ depolymerises giving Monomer and Fluorine compounds.

Decomposition

All Fluorocarbons decompose with flame or red hot metal producing:

HF
HCl
$COCl_2$
COF_2
$COF_2 + H_2O \rightarrow HF + CO_2$ v.v. readily

Intense U.V. light + stable Fluorocarbon $\rightarrow$ Free Cl atoms.

$$Cl + O_3 \rightarrow ClO + O_2$$
$$ClO + O \rightarrow Cl + O_2$$

i.e. $O + O_3 \rightarrow 2O_2$ with Cl regeneration

See W.H.O. Environmental Health No. 6 – 1985.

ACIDIC GASES AND AEROSOLS

e.g.   Chlorine,
Chlorinated gases,
Oxides of Nitrogen,
Oxides of Sulphur.

All are intensely irritant to the respiratory tract.

The higher the solubility, the higher in the R.T. they act.

The absorbed dose is proportional to the partial pressure x solubility x minute volume.

The H^+ Proton destroys cell walls, deeper structures and inactivates enzymes. Oedema, Necrosis and scarring result.

X-Ray changes occur days later, thus they are no use in the early diagnosis.

Irritation – Nose causes brachycardia, brachypnoea, aspiration.

Laryngeal causes cough, spasm, broncho-constriction.

Lung causes brachypnoea, hypotension, brachycardia + occasionally respiratory arrest.

$$Cl_2 + H_2O \rightarrow HOCl \rightarrow H^+Cl^-$$
$$HOCl \rightarrow HCl + O^- \rightarrow H^+ + Cl^-$$
$$COCl_2 + H_2O \rightarrow HCl + CO_2 \rightarrow H^+Cl^-$$

PHOSGENE

$CO + Cl_2 \rightarrow COCl_2$ commercially.

Produced when many Chlorinated Hydrocarbons break down in contact with open flame or hot metal (welding).

e.g. $C\ Cl_4$ in fire extinguishers.

$C\ Cl_2\ C\ Cl_2$ (Tetrachloroethylene) in machining Hi grade steel.

Toxicity

Smelled at 0.125 p.p.m. This is not a safe method of detection.

112

Low-Level Exposure

Initially few symptoms: Mucosal and conjunctival irritation.

Several hours later: Pulmonary Oedema.
Dizziness, freezing feeling, thirst, dyspnoea.
Orthopnoea, cyanosis, frothy sputum.

If the patient survives 2-3 days, usually O.K.

V. occasionally sequellae, but this is unusual and usually affects the respiratory system.

High-Level Exposure

$$CO\ Cl_2\ +\ H_2O \rightarrow HCl + O_2 \rightarrow H^+Cl^-$$

Immediate acid damage to lungs.

Suffocation.

Respiratory arrest.

Treatment:

Supportive.

Prevent chilling.

O^2.

Pulmonary Oedema may need urgent treatment.

OZONE

Ozone is encountered in many industrial situations. It is particularly encountered around the following areas:

X-Ray sources.
U.V. Ray sources.
Electric arcs.
Hg vapour lamps.
Linear accelerators.
Electrical discharges.
Gas shielded welding.

Effects on Exposed People

Irritant to respiratory tract.

Inflammation.

Congestion $\downarrow$ M.E.F.R. This is caused even at levels below the TLV.

Pulmonary Oedema.

Cyanosis.

Haemorrhage – rarely.

Death – rarely.

In mild exposure:

General malaise, headache, dry itchy eyes, tickling throat.

Probably mutagneic.

Method of Causing Lung Damage

Ozone causes necrosis and sloughing of Type 1 pneumocytes in the lung, and there is a great increase in Type 2 pneumocytes.

Impairs activity of alveolar macrophages.

Causes thickening of pulmonary arteries.

Attacks capillary endothelial cells.

Increased infection results from the damaged lung and decreased defence mechanisms.

STYRENE AND POLYSTYRENE

Styrene

$CH_2 : CH_6H_5$ $CH = CH_2$

Ethylene and Benzene $\xrightarrow{\text{Alkylation}}$ Ethylbenzene $\xrightarrow[\text{Dehydrogenation}]{\text{Catalytic}}$ Styrene

Highly reactive. Polymerises readily ($\pm$ with explosion!) Add Inhibitor for transport and storage (Hydroquinone usually).

Use of Styrene Monomer – plastics, additive to resins, synthetic rubber, agricultural compounds, dental fillings.

Styrene in glass reinforced plastic resin systems.

Toxicity

May be inhaled or by percutaneous absorption. Rapidly saturates the body within 30 to 40 minutes.

Low concentrations irritate eyes, mucous membranes, respiratory system, skin.

114

High concentrations:

> Toxic hepatitis, encephalopathy, if major release.
>
> Anaesthesia, drowsiness, vertigo, muscle weakness.

Skin: Dermatitis, secondary defatting and dehydrating. Also, blistering and tissue Necrosis in heavy exposure.

Chronic Toxicity

Central Nervous System.

Upper Respiratory Tract.

Haematological — Leucopoenia but Lymphocytosis.

Hepatic and Biliary Tract.

Ovulation. Epimenorrhoea.

Embryotoxic.

Note — Styrene Oxide:

> Mutagenic.
>
> Reacts with Hepatic cells, Nucleic Acid, Microsomes and Hepatosomal Proteins.

In addition to the above, the Russians report:

> *G.I.T.* irritation, decreased gastric acid, pancreatic changes.
>
> Haematological $\downarrow$ Hb, $\downarrow$ W.B.C., $\downarrow$ Coagulability
>
> Reticulocytosis, $\uparrow$ Capilliary Permeability
>
> Increased Cholinesterase activities.
>
> Increased incidence of sleeping activity/abnormal E.E.G.

Chronic exposure to Styrene alters metabolism of Amino Acids, and thus determination of the Urinary Amino Acid levels can be an indicator of exposure and used for biological monitoring.

Styrene Excretion

85 per cent excreted within 24 hours.

71 per cent in urine as oxidation products of the vinyl group.

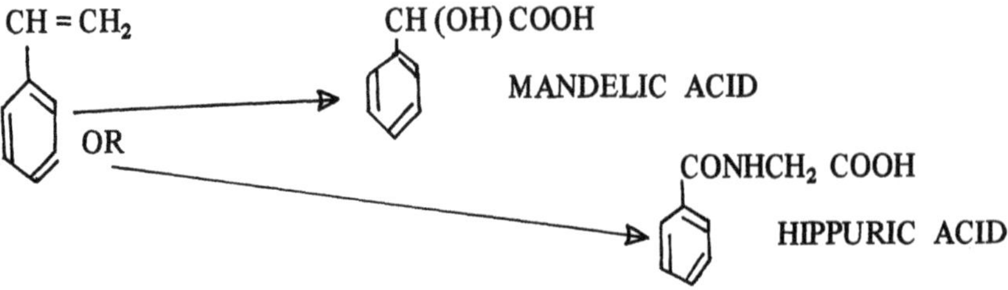

10 per cent in expired air.

Styrene Oxide (more toxic than Styrene).

In high level exposure, Hippuric Acid is the main product.

TLV — Big range. Most countries quote in the range of 50-100 p.p.m., but 1.2 p.p.m. in Russia.

Worker Care

For protection, a worker probably needs air supplied B.A. whilst doing big mouldings.

Exclude people with target organ abnormality.

Exclude pregnant women.

Monitor regularly.

Polystyrene

Polymerisation is a strongly exothermic reaction and can cause explosions.

Copolymerisation of Styrene + other Monomers,

    e.g. Methyl Methacrylate.
          Methyl Styrene.
          Acrylonitrile.

All ingredients highly flammable.

Polystyrene readily ignites (160-260°C).

Foamed Polystyrene is very toxic when burned.

Foamed Polystyrene production may produce vapours: Styrene;
Isopentane;
Benzaldehyde;

A.B.S. (Acrylonitrile/Butadeine/Styrene).

A.B.S. Styrene Copolymer production may produce vapours: Styrene;
Divinyl;
Acrylonitrile;
Benzaldehyde;
HCN;
CO;
Mercaptans.

These various compounds added in manufacture may be the cause of serious toxicity during a fire.

Air in the premises may also contain Styrene Dioxide and Acetophenone.

Toxicity depends on Copolymers involved.

Most toxic are: Styrene;
Acrylonitrile;
Methylmethacrylate.

Exposure Affects Several Systems

Central Nervous System — E.E.G. change — Coma — Encephalopathy.

Blood system — Leucopoenia — Mutagenesis.

Liver — Toxic Hepatitis.

Reproductive system in women — Pregnancy — Toxaemia, Placental change, Epimenorrhoea.

May also be involved in Carcinogenesis.

MERCURY AND SALTS

Uses: Thermometers.
Electrical switches + batteries.
Felts — carrotting.
Paints.
Explosives.
Organic for fungicides.
Various medically (Mersalyll).

Action

Hg. Vapour — Lipophyllic $\rightarrow$ Brain $+ \; O_2 \; \rightarrow \; Hg^{++}$ (toxic)

Hg. reacts with Sulphydryl groupings.

Acute Toxicity

Ingestion: Gut-burning, sloughing of intestine.
                 Bloody diarrhoea.
                 Kidney — Anuria may lead to death.

Inhalation: Respiratory System affected.
                 Central Nervous System (especially Alkyl Mercury). Permanent changes developing over years are irreversible.

Chronic

Varied — Gut/liver/kidney/central nervous system.

Mercurialism.

Blood levels > 0.2 μgm/Ml Methyl Hg causes symptoms.
                   < 0.1 μgm/ml Organic Hg causes symptoms.
                   < 0.1 μgm/ml Inorganic Hg is neuromuscular toxic.

Urine < 0.3. Hg/24 hours probably poisoning.
       Group average > 0.1 mg Hg/24 hours = work problem.
       Individual > 0.2. mg Hg/24 hours remove till < 0.05 mg.

Organic   Remove if excretion > 0.05 mg Hg/24 hours.

Treatment of Mercury Intoxication.

1.   Gastric lavage.

2.   B.A.L. (Dimercaprol) for 1/52

$$H_2C - CH_2 - CH_2OH + Hg \qquad\qquad H_2C - CH_2 - CH_2OH$$
$$|\qquad\quad|\qquad\qquad\qquad\qquad\qquad\quad |\qquad\quad\ |$$
$$SH\quad\ \ SH\qquad\qquad\qquad\qquad\qquad S\diagdown\ \ \ \diagup S$$
$$Hg$$

3.   Renal Dialysis, Anuria.

4.   Penicillamine.

5.   Gut damage, including perforation.

CHROMIUM

Main uses: Steel alloys.
                Refractories.
                Printing and engraving.
                Tanning, leather.
                Dyes and pigments.
                Electroplating and anodising.
                Cement.

Industrially occurs as Liquid.
Dust.
Vapour (especially welding and oxy-cutting).
Metallic (rare) (Important in the development of cancer with asbestos).

Physiology: Essential for Glucose Metabolism

Valency: Hexavalent and Trivalent important occupationally.
Hexavalent toxic.
Trivalent probably safe and is the form in which it occurs in the body.

Effects

Mainly skin (reported 1827) and respiratory system (1863).

R.T. — Hexavalent   Painless nasal perforation.
Pneumoconiosis — possibly.
Sensitisation of respiratory tract.
Pneumonitis, Bronchitis $\rightarrow$ Ca Lung
(may be 20 years latent period).

Hexavalent dust ⎫
Monochromates ⎬ Dose-related cancer cause.
Bichromates ⎭

Skin — Ulcer
Non-ulcerative Dermatitis.
Allergic contact Dermatitis.
Sensitisation.

"Chrome Hole" is an ulcer formed in cracked skin. Heals v.v. slowly and leaves a scar. Caused by Chromic Acid, Na, K, & NH_4 Dichromate and Na, K Chromate.

Protection

10 per cent Calcium E.D.T.A. cream used on hands, especially in the nostrils.

Clothing should be impervious to solutions.

Monitor

Monitor Air sampling

Cannot monitor blood or urine.

Regular inspection (2/52) to look for signs of nasal perforation or effects on the skin.

CYANIDES

Uses: Fumigant.
Chemical synthesis.
Metal extraction from ores.
Metal cleaning.
Electroplating.

T.L.V.   O.H.S.A.   5 mg/m^3 /8 hour.
N.I.O.S.H.   5 mg/m^3 /10 minutes.

Action: Interferes with Cytochrome Oxidase, an essential enzyme in oxygen transport, and thus

Prevents oxygen transport to the tissues and its release in the tissues.

Fatal Dose $\simeq$ 0.5 p.p.m. HCN.

Storage: N.B. Story dry as $XCN + H_2O \rightarrow HCN \uparrow \quad + CO \uparrow$

Acute Poisoning

$<$ 1 M.L.D. causes   Dizziness.
Flushing.
Tachycardia.
Hypotension.
Drowsiness $\rightarrow$ Coma $\rightarrow$ Death within 4 hours.

Percutaneous absorption can cause these symptoms.

Note: Sodium Nitroprusside kills in about 12 hours.

Chronic Poisoning

Dizziness, flushing, loss of weight,
Tachycardia hypotension,
Hypothyroidism,
Mental deterioration

Acrylonitrile causes Ca. bladder
Laetrile causes Agranulocytosis

Treatment

Simple   —   Oxygen — artifical respiration using O_2 should be available.

—   Amyl Nitrite every 5 minutes (unless B.P. $<$ 80 mm) — glass ampoules broken and the vapours inhaled.

Oral   "A"   15.8% Ferrous Sulphate.
0.3% Citric Acid in distrilled water.

"B"   6.0% Sodium Carbonate.

Swallow 50 ml of each solution. This may be given awaiting medical help.

Medical (a) $-$ I.V. 3% Sodium Nitrite $+$ Hb $\rightarrow$ MetHb $+$ Cn$^-$

MetHb $+$ CN$^-$ $\rightarrow$ Cyanmethaemoglobin   (Keep $< 40\%$)

then $-$ I.V. Sodium Thiosulphate $+$ CN $\rightarrow$ Thiocyanate

  (b) $-$ Dicobalt Edetate $-$ Note may be lethal if no CN. It is often safer to await arrival in hospital and testing to ensure that cyanide poisoning has occurred.

VANADIUM

Occupationally occurs in: Fossil fuels (coals and oils).
Steel, (Ferrovanadium and Vanadium Carbide).
Dyes.
Ceramic Glazes.
As a catalyst in many processes.

Absorption: Inhalation
Ingestion
Percutaneous

Toxic Effects

(a) Respiratory tract: Rhinitis.
Sore throat.
Persistent cough.
Non-infected bronchitis.
Lung cancer (often 20 years latent period before it develops).

Note: Wheezing and severe Dyspnoea may occur within 6-24 hours.

(b) Skin $-$ Eczema.
Nail composition altered.

(c) General $-$ Malaise.
Headache.
Loss of weight.
Tremor.

Monitoring $-$ Urine $-$ 60 per cent excreted within 24 hours.

$-$ Blood $-$ technically not an easy assay, but very accurate.

$-$ Nail clippings $-$ Cystine content falls in chronic *low* grade exposure.

— Annual lung function test and full medical may show early signs of deteriorating pulmonary function.

— Periodic chest X-ray.

TLV For Pentavalent dust 0.5 mg/m^3 V_2O_5

Fumes 0.05 mg/m^3

Note: Trivalent less toxic than Pentavalent

DISASTER PLANNING AND CHEMICAL ACCIDENTS

by
Peter Hollingworth

Chemical accidents are not a rare occurrence in industry. They should not be permitted to develop into disasters. When considering several of the recent disasters associated with chemical accidents, there was a stage, after the initial incident, where, if the correct action had been taken, the disaster would have either not occurred or would have been less serious. In these cases no one had developed a plan to cope with such accidents and prevent them from turning into disasters.

The *PLAN* is all important, but it must be a *PRACTICAL* plan, widely circulated or *PROMULGATED.* Importantly, it must be frequently *PRACTISED* and, following all practices or emergencies, a debriefing session may allow *PERFECTING* or modifying of the plan, which then recommences this continuous cycle.

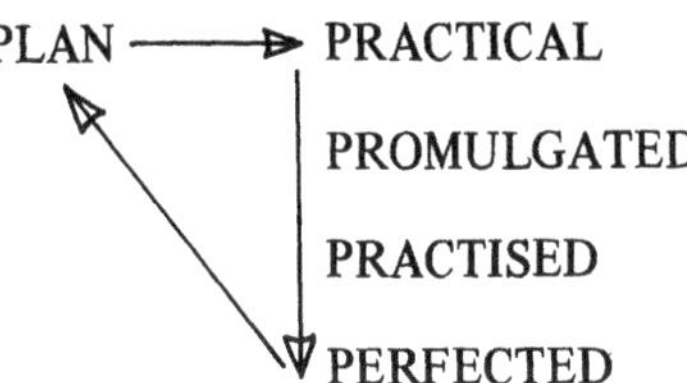

PLAN — Stages in Preparation:

    (a) Define *Problem.*

    (b) Assess *Impact* on surroundings.

    (c) Analyse *Requirements.*

    (d) Assess *Resources* required to meet (c) and those currently available to meet (c).

    (e) Establish *Liaison* with other bodies.

    (f) Establish *Co-ordinating* body.

    (g) Produce *Plans,* both *General* and *Contingency.*

(a) DEFINE PROBLEM

This stage involves the preparation of several scenarios. Some will involve detailed knowledge of the chemical process and the potential aberrations which may occur under various conditions. Other scenarios will involve external accidents which may affect the work site. These may be of a chemical nature, e.g. exploding tanker of L.P.G.; of a physical nature, e.g. fire, flood, riot, earthquake; or due to rare incidents, such as crash of an aeroplane. All the scenarios will require different responses from the workplace as the hazards posed will vary, often very significantly, according to the threat to the workplace.

Frequently, there will not be the resources within the workplace organisation to provide all the technical answers as to the best methods to cope with the hazards.

(b)   ASSESS IMPACT ON SURROUNDINGS

An accident in a chemical plant will not only pose a threat to those who are working there, but will also pose a threat to the surroundings. Such threats may be of fire, spread of toxic gases or spill of toxic flammable liquids. The area threatened may be just proximate to the site of the chemical plant or may be very widespread. An extreme example of an industrial accident would be the Chernobyl nuclear accident or the release of chemicals into the Rhine in Switzerland and the flushing down through several countries till the contaminated water reached the North Sea.

When assessing the impact, it is important to try and assess what are the potential hazards already near the chemical plant. It may be possible for a chain reaction to occur where a series of workplaces all add to the disaster potential if one is subject to an accident. This aspect will be further examined under (e).

In assessing the impact, the time element is important in several different ways. Thus, the impact in a commercial area at 2 a.m. may be quite different from that at 2 p.m. In an area where the chemical plant is surrounded by high density population, the time of day at which the accident occurs will affect the size of the population at risk and needs to be taken into account when planning a response. In some less developed areas, the response from distant specialist helping agencies may be different, according to daylight or weather conditions.

Remember that in nearly all chemical accidents which became disasters, the majority of the victims had nothing to do with the chemical plant, and certainly they were not workers there.

(c)   ANALYSE REQUIREMENTS

There are two quite distinct ways in which the analyses must be carried out:

First, the requirements must be set out in chronological order from the time of the accident:

(i)   Immediate: This period usually covers the initial minutes, perhaps as long as two hours, depending on the nature of the accident.

124

(ii)   Short-term: This covers early requirements and frequently involves special-
                        ist response usually measured in hours, less than 24.

(iii)   Medium-term: Depending on the nature of the accident, there may exist a
                        hazard over several days and a need to provide a response to
                        contain this hazard.

(iv)   Long-term: This will certainly involve specialist groups in long-term
                        decontamination or rehabilitation. Thus, areas around Seveso
                        in N. Italy can still not have crops grown there several years
                        after a chemical accident developed into a disaster.

Second, the following requirements should be considered:

(i)   Human: Specific skills.
                        Where normally located.
                        Numbers.
                        Availability   — response time.
                                          — duration.
                        Level of training.
                        Local support needed.

(ii)   Equipment: General or specific.
                        Capability.
                        Availability   — normal use.
                                          — where kept.
                                          — operators needed
                                            (sophistication).
                                          — durability.

(iii)   Special Facilities, e.g. Hospitals
                        Detoxification units

Remember that in preventing an accident developing into a disaster, the most precious commodity is TIME. When considering the Human, Equipment and Special Facilities' requirements, it is essential to carefully consider TIME. Where are they located? How long does it take to contact them or have them on site?

Remember that in the case of chemical disasters there are frequently only very few human resources with practical experience of coping with such disasters. They may not even be in the country in which the accident occurs. Good planning will have identified them and contacted them well in advance, and obtained much counselling and help long before the accident occurs. It would then be possible to advise from thousands of miles away with a detailed knowledge of the workplace.

(d)   ASSESS RESOURCES

This is the important stage of *quantifying* requirements and is a difficult stage. It is obviously as potentially dangerous to have too much unwanted equipment as it is to have either inappropriate or insufficient equipment.

It must also be remembered that sophisticated equipment require operators who are trained in its use. Just because a person is capable of driving a car does not mean that he can cope with a mobile crane! So, in many cases the specialist resources will have both human and mechanical components. Sophisticated equipment, without operators, is useless.

Frequently in chemical accidents there are large numbers of people affected. The numbers can be mind-boggling and will easily swamp normal resources. Thus, in the L.P.G. explosion in Mexico City there were approximately 8,000 people involved — 500 died, 900 were admitted to over 50 hospitals, nearly 2,000 were seen as hospital out-patients and about 5,000 were seen in emergency medical centres after the hospital resources had been swamped.

Particularly in chemical accidents, it is of paramount importance to plan resources on a time-basis. Some chemicals, whilst causing minor irritations of the eyes, nose or skin, as immediate side-effects, may not produce pulmonary complications until 24 or 48 hours later. It is essential to make allowances for this second "wave" of victims when planning both human resources and facilities.

(e)   ESTABLISH LIAISON WITH OTHER BODIES

Depending on the accident envisaged, there may be only need to involve *local emergency services,* such as the police, fire and ambulance. In other cases, *essential services,* such as gas, electricity and sewerage, will be involved. Occasionally, military or paramilitary help is needed in evacuation, transport, special equipment and operators, security and other services. Depending on the location of the emergency, there may be involvement, either as potentially threatened or as assistance, of mass transport, ferries, airports and port authorities. It is too late to attempt to establish liaison with such bodies *after* the accident. They must be involved at the planning stage.

(f)   ESTABLISH CO-ORDINATING BODY

This follows from the previous paragraph. It is too late to untangle chains of command after the accident. In many cases there will need to be high-level national or State permission for some of these agencies to be subjected, even in emergencies, to the command of another Government Minister or Department. Occasionally, there will be need for the Government to pass legislation establishing such a *Co-ordinating Body,* assigning special *Emergency Powers* and laying down the administrative pathway whereby a *State of Emergency will be proclaimed.* If life and property-threatening delay is not to occur after the accident, all this structure and mechanism should be established well in advance.

(g)   GENERAL AND CONTINGENCY PLANS

It will always be possible to prepare general plans for such common occurrences as fires, or, in this part of the world, floods, earthquakes and typhoons/tropical cyclones. Such plans must, of necessity, be flexible and will only have a general idea of what counter-disaster steps will be taken.

In the chemical industry, in particular, very detailed contingency plans can be prepared, depending on the possible accidents which can occur in the process. All possible undesirable chemical reactions and unwanted intermediate or by-products can be considered. Where part of the response would be to use chemical agents, either as neutralisers or inactivators, detailed planning is possible concerning the nature, quantities, storage and use.

All of these plans need wide *promulgation,* to be *practised* regularly, especially if there is a rapid turnover of the workforce, and need to be constantly updated as the plant or process is modified.

COMMUNICATIONS

In all emergency situations it is essential to have good communications. In the physical sense, it is important to realise that communication is a specialist function.

Within the *Co-ordinating Body* it is important to set up a dedicated *Communications Group* with the necessary specialist operators and equipment.

Counter-Disaster Operations depend on *Essential Communication Links.*

Essential Communication Links may take any of the following forms:

(a)   Human communication: runners, messengers, despatch riders, couriers, loud-speakers.

(b)   Audio-visual: sirens, bells, flags, lights (semaphore), notice boards, chalk, paint.

(c)   Cable: fixed telephones or emergency field telephones.

(d)   Two-way radios.

(e)   Electronic media: radios, television.

Accurate COMMUNICATION of the details of a disaster is critical to the Co-ordinating Counter-Disaster Body's correct response. Several components are vital and need regular updating as the operations develop and the incident continues.

(a)   Definition of problem:

(i)   Specific chemical nature of incident includes order of magnitude of amounts of chemicals involved, as well as state of reaction of people and effects on them. This is particularly important where the release of toxic chemicals may imperil the emergency response teams.

(ii)   General advice on local factors, such as nature of surrounding industry, hazards, population.

(iii)   Weather conditions, especially relating to wind conditions. This information is critical both from the point of view of evacuation and safeguarding emergency response teams.

(b)   Assessment of immediate impact on surroundings.

(c)   Analysis of requirements, especially if unusual specialist needs can be accurately foreseen.

(d)   Assessment of resources: current resources on site and state of deployment, early warning to areas with long response time.

(e)   Liaison: Has any direct contact already been made with other resources required? (fire, police, ambulance, etc.)

Name current person directing Counter-Disaster Activities.

Name person and location of Site Liaison Officer who should remain at the end of the line of communications.

The Co-ordinating Body may need to liaise with national or international authorities.

MASS CASUALTY PROCEDURES

Mention has already been made of the very large numbers that may be involved. Unless there has been forward-planning, chaos can result.

Command

Needs establishment prior to incident.

Needs urgent arrival of local commander on site after incident.

Needs high profile. Ready identifiability of local commander. There may be the need to visibly identify by special uniform or coloured clothing.

Control

Local commander needs *Site Control.*

H.Q. needs *Overall Control.* There may be several sites to be controlled and co-ordinated by the H.Q.

Communication

H.Q. to *Site*

H.Q. to *Response Personnel.* There are four stages to mobilisation of personnel:
    alert,
    stand-by,
    call-out; and
    stand-down.

H.Q. to *Response Organisations:* fire, police, ambulance.

H.Q. to *Community Support:* essential services.
                             catering/shelter/welfare.
                             public health.
                             drinking water.
                             transport.
                             religious orders/undertakers.

H.Q. to *Potentially-Threatened Services:* rail, air, ferry, sewerage, water.

H.Q. to *Media.*

Site Control to other *Site Locations* and to *H.Q.*

"**Triage**" is the French word for "sorting". The concept was introduced by French surgeons in the Napoleonic wars to prevent squandering of scarce resources on the unsalvageable.

Triage: Greatest good for greatest number.
         Evaluation and classification of casualties for the purpose of evacuation/treatment in the most appropriate place.

         Need for advice *to doctors* on chemical hazards. Few doctors will have a ready understanding of potential chemical hazards, and so need expert advice.

Triage into 4 groups: (a) Need stabilisation prior to evacuation.

                       (b) Fit for immediate evacuation.

(c) With minimal treatment, may remain — these could include plant workers with a detailed knowledge of the workplace and process, who would assist emergency teams.

(d) Beyond treatment — need pain relief or comforting.

Medical Needs

The medical needs to respond to a chemical accident are dealt with in this section. As part of the plan development, they should be involved at an early stage so that they may fully understand the nature of the chemical hazard and prepare appropriate counter-measures.

Personnel

Primary Disaster Response Team: Surgeon.
Anaesthetist.
Nurses.

Follow-up Team: Rarely needed. Manned as situation demands.

Volunteers: Blood donors.
Religious orders.
Crowd control.
Social workers/psychologists.
Para-medical professionals.
Interpreters
Drivers (cars, taxis, buses).
Onlookers may help with the temporarily blinded by leading them to treatment areas; helping console or protect injured from the elements.

Penciller(s) for Triage Officer: Records must be kept, particularly of where injured people are dispatched to.

Equipment

Personal protective equipment for medical team in essential.

Identification of medical team.

Portable packs, waterproof and flame-proof:

(a) life support/resuscitation;
(b) painkillers;
(c) splints;
(d) dressings; and
(e) markers/records.

130

Emergency lighting, generators.

Specialised stretchers.

Pharmaceutical supplies.

Transport

Ambulances (only where specialist transport needed).

Cars/taxis (suitable for many evacuations).

Buses (often needed for "walking wounded").

Airborne/helicopters.

Hospital(s)

"Triage-at-Door". Contaminated patients segregated.
Beds, theatres, X-ray, laboratories.

Relatives, staff, telephonists, catering, stores, pharmacy, will all need mobilisation or freeing-up to respond to the emergency. For several there is a "lead time" before they become available. Early warning after the initial communication of the accident permits appropriate procedures to begin to be implemented.

Long-Term Needs

Psychiatric problems may persist for years after a major disaster.

Monitoring exposed populations.

Monitoring of clean-up teams, emergency teams and detoxification unit's personnel.

Public Health

Contamination of water, food supplies.

Contamination of food-chain.

Sanitation (sewerage, pest/insect control, dead disposal, animal disposal).

Immunisation against infectious diseases, i.e. cholera, typhoid, typhus.

Others

Although this is the last mentioned, it is most important.

Pre-incident, Post-incident and On-going:
    Training, Training, Training!
    Immunisation of staff who may be called upon.
    Debriefing after event or practice.
    Must include "dummy runs".
    Familiarisation trips around high-risk areas.
    Monitor staff for exposure.

IMMEDIATE MEDICAL CONSEQUENCES OF CHEMICAL ACCIDENTS

There are several types of chemical accidents.

Blast from BLEVE (Boiling Liquid Expanding Vapour Explosions)
or Fuel/Air, Vapour Phase Explosions.

The effects on gas-filled cavities occur when the gas attempts to expand, i.e. in the explosion phase.
Solid organs are less sensitive.
Eyes.
Ears — Trauma + Noise: frequently survivors have temporary or permanent deafness, following a blast.

Toxic Gases

(a)   Pose greater threat than flammable cloud, and there will be need to monitor the population over a wider area.

(b)   Hazards to fixed medical installations, e.g. hospitals, which may need evacuation.

(c)   Emergency medical team evacuation may be required, unless they all have adequate protection.

(d)   There is need for constant monitoring of site and people.

(e)   There is need for meteorological data as wind-shift can alter the population at risk. Rain may react with exposed dry chemicals.

(f)   Downwind vapour cloud dispersion.

(g)   There is need for awareness of *delayed* symptoms, which may not be present for 1 or 2 days with several toxic gases.

Fire

Primary

Foreign bodies in eyes may occur over a wide area or period. Numbers may be enormous. Treatment is usually simple and does not need specialist care.

Skin — heat burns, chemical burns, soot.

Respiratory tract irritation with heat, smoke or toxic fumes.

Secondary

Due to fire-fighters and the fire-fighting agents which they use.

(a)   Halogenated HCs and Halons in extinguishers may degrade to Phosgene and Chlorine — not too commonly used now.

(b) H_2O may liberate toxic gases, e.g. HCN with Acrylonitrile or Acetone Cyano-hydrin.

(c) H_2O may produce caustic solutions with NH_3, Cl_2, NaOH.

(d) H_2O may produce acid solutions, e.g. with Nitrogen Tetroxide.

(e)   Unaware of degradation products, which may be toxic.

(f)   Unaware of signs and symptoms of toxicity, so fire-fighters may be over-come whilst fighting the fire.

Aircraft

Most fatalties in commercial airline crashes are due to Cyanide Poisoning from the burning of plastics or foams, or from smoke inhalation.

Radiation

Radioactive gas clouds may be released. There are the dual medical problems of *acute toxicity* for both the emergency teams and the population at large, as well as *long-term monitoring* needs for the whole area and population.

Natural Phenomenon

Wind, temperature : Exposure victims and rescue team all can be affected;
                       : Delays in transport, arrival of rescue teams due to adverse weather conditions.

LEGISLATION AND EXPOSURE STANDARDS
FOR CHEMICAL SUBSTANCES
by
Barry Chesson

INTRODUCTION

There was a recognition long ago that there exists a relationship between the environment and workers' health. The biblical reference: "He that diggeth a pit shall fall into it: and whoso breaketh a hedge, a serpent shall bite him. He that cleaveth wood shall be endangered thereby", is derived from an unknown writer's observations more than 5,000 years ago. In antiquity, little was done to protect workmen in "dangerous trades" because they were considered to be expendable. In many civilisations the ready supply of slaves provided an obvious solution to the labour turnover problem. Over the past two millennia certain key individuals have emerged to draw attention to the need to protect workmen from hazards associated with their occupations (Appendix I). However, it has been only in comparatively recent times that there has been any significant effort to adopt preventive practices.

State intervention in the occupational health and safety area dates from the period of the industrial revolution in Britain, when working conditions gave rise to high incidences of injury and ill health. Early statutes were directed at the textile industry, but were largely ineffective since few funds were appropriated. The Shaftsbury legislation of 1833 was significant in that it extended control beyond the textile industry for the first time, and also because it provided for a factory inspectorate with powers of enforcement, backed by legal sanction.

Subsequent legislative activity in this area was characterised by a proliferation of detailed legislation, interrupted periodically by attempts at consolidation. It was generally believed that a highly prescriptive approach was the best way to ensure safety in the workplace.

MODERN LEGISLATIVE APPROACHES

The legislative approach depends for its success on three factors. First, the legislative requirements should be relevant to the injury causes that they purport to control. Second, legislative requirements should be capable of implementation, that is, they should be readily available, easy to comprehend, technically practicable, and permit ready detection of non-compliance. Finally, the legislative system should contribute to the development of expertise among all those persons whose occupations are concerned with health and safety controls.

Many countries have recently reviewed and remodelled their legislative framework for occupational health and safety. In the United Kingdom, in 1970, a committee of enquiry was set up to assess the shortcomings of existing safety and health legislation. Its report, commonly known as the Robens Report, was presented to Parliament in 1972 and gave rise to the Health and Safety at Work Act, 1974. The modern British approach to control of workplace hazards is based on five principles:

(a) It is the employer's responsibility to prevent health risks arising from his plant and processes.

(b) The worker has a right to know the risks and has a duty to co-operate with his employer in controlling them.

(c) Standards for control are developed through consultation between industry and labour representatives.

(d) The remedies should be proportional and appropriate to the risks.

(e) A high standard of compliance with legal requirements is expected and enforced by an expert inspectorate with strong powers.

The HASAWA allocated responsibility to everyone concerned with health and safety at work — employers, employees, self-employed, manufacturers and suppliers of industrial equipment and materials.

Employers are required to maintain an up-to-date written statement of the safety policy and details of the arrangements that they have made for its implementation.

Encouragement is given to the setting up of "in-plant" or "in-house" committees which structure rules applicable to their particular establishments. Under the Act, regulations may be made describing the circumstances for the election or appointment of safety representatives from employee ranks.

Rules produced in this way by the in-plant committee are vetted by the Health and Safety Commission and are then administered on a day-to-day basis by the local group. Overall supervision of the legislation, however, rests with inspectors employed by the Commission.

Inspectors pay close attention to the "basic-approach concept", and their training is now including greater emphasis on concepts, philosophies and basic principles.

A feature of the Act is that it provides safety inspectors with power to issue prohibition notices and "on-the spot" fines, thereby obviating the previous requirement of prosecution through the courts.

A number of countries have chosen to adopt Robens-style legislation. In Sweden, for example, legislation emphasises a joint management-employee approach in protecting health and promoting accident prevention. Statutes are supplemented by voluntary agreements between the relevant parties. Sweden's Work Environment Act provides for the appointment of safety representatives and the setting up of joint safety committees.

In the United States, the Occupational Safety and Health Act of 1970 was a piece of landmark legislation. In common with the U.K. approach, it provided a framework for the application of detailed standards and for the inspection and enforcement of such provisions. It also defined the general duties of employers and employees. However, in contrast to the British legislation, which opted for a self-regulating approach, the United States has adopted a prescriptive one, and their provisions for employee participation in achieving safety and health improvements in the workplace are more limited. The main requirements of the Occupational Safety and Health Act (1970) are as follows:

(a)   Occupational safety and health standards be developed and promulgated, initially using national consensus standards and established federal standards, then adding emergency standards and standards developed through a prescribed procedure outlined in the Act.

(b)   Federal action be taken to enforce the standards, at least for the initial period of the Act. When a State develops and enforces a job safety and health programme, which is at least as effective as that developed by OSHA under the Act, it may assume exclusive jurisdiction over the health and safety conditions of employees within the State.

(c)   The necessary research be conducted to provide the information needed to set and enforce proper standards.

(d)   A training and education programme be established to develop expertise in the field and to produce a corps of safety and industrial health manpower.

EXPOSURE STANDARDS

Concomitant with initiatives to reform occupational health and safety legislation has been a drive to develop permissible limits of exposure. Exposure limits are the commonest form of health and safety standards applied to toxic agents in an occupational setting.

Maximum Acceptable Concentrations (MACs)

MACs emanate from the USSR. They are exposure standards based on the philosophy that it is desirable to limit occupational exposures to levels that produce no measurable effect, even with the most sensitive detection systems available. The practicability of controlling exposures in the workplace to the MAC criteria is not a factor in establishing such levels. MACs are usually seen as long-term goals.

Threshold Limit Values (TLVs)

The most widely used set of numerical values for limiting exposure to workplace contaminants is the list of "Threshold Limit Values for Chemical Substances in the Work Environment" published annually by the American Conference of Governmental Industrial Hygienists (ACGIH). The list includes values for about six hundred of the chemicals most widely used in industry and for a number of physical agents. According to the ACGIH, values are derived from the best available information from industrial experience, from experimental human and animal studies and, when possible, from a combination of all three.

Threshold Limit Values represent airborne concentrations to which nearly all workers may be repeatedly exposed day after day without adverse effects. Because of wide variations in individual susceptibility, however, a small percentage of workers may experience discomfort from some substances at concentrations at or below the threshold limit; a smaller percentage may be affected more seriously by aggravation of a pre-existing condition or by the development of an occupational illness.

The ACGIH strongly emphasises that their limits should not be taken as fine lines between safe and dangerous, and that the best practice is to maintain concentrations of all atmospheric contaminants as low as is practical.

Three types of TLV have been developed by the ACGIH — to allow for time as well as concentration as a factor in exposure:

(a) The TLV-TWA: the time-weighted average concentration for a normal 8-hour work-day and 40-hour work-week, to which nearly all workers may be exposed, day after day, without adverse effects.

(b) The TLV-STEL: a time-weighted average value calculated over a 15 minute period. This is applied in situations where brief high excursions could be experienced (while not exceeding the 8 hour TLV-TWA).

(c) The TLV-C: a value which should not be exceeded evenly briefly. It is used in situations where acute effects might be experienced — as with sensitisers, irritants and other quick-acting materials.

If any one of the three TLVs is exceeded, a potential hazard from that substance is presumed to exist. Some listed substances are followed by the notation "skin", which refers to the potential for exposure to occur via the skin, mucous membranes or eyes, either by airborne means or by direct contact with the substance. Biological monitoring may be appropriate in determining the contribution of such exposure to the total dose.

MIXED EXPOSURES

Exposure to a single chemical is an exception rather than the rule. There are many situations in industry where employees are exposed to a mix of chemical (and physical) agents during the course of their normal activities. In addition to these exposures, employees may absorb other foreign compounds (such as alcohol, drugs and tobacco smoke and its combustion products) outside the working area.

Each step in the metabolic pathway of a chemical in an organism can be influenced by another foreign compound or its metabolites, producing a toxicokinetic interaction. The combined effects arising out of multiple exposure may be synergistic (enhanced or potentiating effect), antagonistic (reduced effect), additive or independent in nature. Discussion on the application of TLVs to mixtures is provided in the TLV booklet.

NON-TRADITIONAL WORK SCHEDULES

A further difficulty in applying TLVs occurs when work schedules are outside of the 8-hour-per-day, 5-day-per-week routine. Some shift rosters used in industry operate on a 12-hour work-day basis (and alternate between a three- and four-day work-week; others are based on seven continuous 8-hour days, with 2-4 days off between shift changes.

It is possible that prolonged exposures to such schedules might give rise to adverse health effects. In a 12-hour work-day, the period of exposure is 50 per cent greater than an 8-hour work-day, and the period of recovery is reduced by 25 per cent. The higher body burdens of a contaminant might stress the body's detoxification and clearance mechanisms to a point where accumulation of a chemical reaches harmful levels.

The technical literature contains a number of sophisticated guidelines for adjusting permissible limits.

However, where the exposure limit is intended to control nuisance conditions or odour, no modification is needed to protect employees on non-traditional work schedules. The same generalisation applies to all air contaminants assigned a ceiling limit.

Modified exposure limits may be necessary for the remaining air contaminants where exposure limits are based upon systemic effects, narcosis and other acute and chronic health effects. Many factors are considered in determining the need for a reduced exposure limit, including the metabolism and biological half-life of the material.

It is generally held that it is unnecessarily restrictive to reduce limits simply in proportion to the extended work-day or work-week. Each situation should be reviewed individually and recommendations made on a case-by-case basis.

REFERENCES:

1. Commonwealth Department of Science and Technology (1981). Legislation at Work: Safety and Health. Australian Government Publishing Service, Canberra.

2. Clayton, G.D. and Clayton, F.E. (1978). Patty's Industrial Hygiene and Toxicology. 3rd Edn, Vol. 1. John Wiley and Sons, New York.

3. United States Government (1970). Occupational Safety and Health Act. Washington.

4. Robens et al (1972). Safety and Health at Work — Report of the Committee, 1970-1972. Her Majesty's Stationery Office, London.

5. Worksafe Austalia (1986). Occupational Health and Safety. Australian Medical Association, Canberra.

6. ACGIH (1985). Threshold Limit Values and Biological Exposure Indices for 1985-1986. American Conference of Governmental Industrial Hygienists, Cincinnati, Ohio.

Appendix I

HISTORICAL BACKGROUND

The word "hygiene" may have come from the name of the Greek Goddess of health — *Hygeia.*

(400 BC)	*Hippocrates* first reported occupational diseases associated with slaves who worked in lead mines.
(23-79 AD)	*Pliny the Elder* — The Roman scientist wrote about dangers of zinc, lead, mercury, sulphur, and first described a mask made of a pigs bladder which could be used to protect against dust.
(250 AD)	*Galen* pinpointed the danger of acid mist in copper mines.
(1500's)	*Agricola,* a German, published De Re Metallica, a 12-volume series on the hazards of mining. He recommended ventilation and masks to protect against dust. *Ulrich Ellenbog* and *Theo Bombastus von Hohenheim (Paracelsus)* also published work on occupational diseases in this same century.
(1700's)	*Bernardo Ramazzini* published De Morbis Artificum dealing with the pathology of silicosis found in autopsies of miners — often called the Father of Occupational Medicine, an Italian Physician.
(1802)	Health and Morals of Apprentices Act (U.K.) — the first instance of State intervention in private enterprise.
(1833)	Passage of the first British Factories Act — it provided for the appointment of inspectors and levied substantial fines for contraventions.
(1864)	British Factories Act — contained the first requirements for ventilation.
(1901)	British Factories Act — provided for the making of regulations or orders to control dangerous trades.
(1908)	First U.S. Federal Compensation Act covering selected civil service employees.

(1936)   Walsh Healy Public Contracts Act (U.S.A.) mandated safety/health requirements for Government contractors, precursor to the Occupational Safety and Health Act.

(1970)   Occupational Safety and Health Act (U.S.A.).

(1974)   Health and Safety at Work Act (U.K.).

SAMPLING STRATEGIES AND TECHNIQUES FOR CONTAMINANT EVALUATION

by
Barry Chesson

INTRODUCTION

Over the past few years there has been a remarkable surge in awareness and interest in occupational health and safety issues. Governments, employers, unions, teritiary institutions, media groups and the wider community have recognised that the link between the workplace environment and the wellbeing of the employee deserves attention. Increasingly, industrial organisations are realising that this aspect of management of human resources may have significant ramifications in terms of the profitability and viability of the operations.

Against the backdrop of this recent activity, the formerly obscure professional pursuit of industrial (or occupational) hygiene has achieved some prominence. Industrial hygiene is often defined as the recognition, evaluation and control of physical, chemical and biological hazards in the workplace. Physical hazards may be presented by agents, such as noise, vibration, radiation and thermal stress. Chemical hazards may result from the presence of agents, such as dusts, fibres, gases and vapours. Biological hazards are less important but may arise in industries or occupations where there is potential for contact with parasites, fungi or bacteria.

This paper will focus mainly on procedures for recognition and evaluation of airborne contaminants (chemical hazards). However, many of the principles outlined have wider application.

PRELIMINARY INDUSTRIAL HYGIENE SURVEY

A basic, systematic procedure should be used to establish whether workplace conditions are likely to give rise to adverse health effects.

The first step usually involves conducting a preliminary survey of processes, work operations and job procedures, together with the setting up of an initial inventory of potential hazards and control systems that are in use. It is important to consider raw materials, intermediates, products and by-products associated with the process. Each step must be examined to determine hazards associated with normal activity, abnormal operating conditions and possible emergency situations.

It is necessary to understand the process well enough to establish where air contaminants are produced and where and for how long employees are exposed. In addition, when seeking to identify possible hazards from processes and materials, the potential effect on the general environment in the vicinity of the operation should be taken into account.

The second step is to conduct a walk-through survey of the site, to establish the accuracy of the information that has been obtained per the above. Sensory preception plays an important role in this activity. The sense of vision can be used to establish sources of particulate generation, the state of housekeeping in the area, deposition of contaminants on the face or clothing of workers and many other indicators of potential problems. Also, the presence of many gases and vapours can be detected by odour or irritation. Note should be taken of how ventilation devices, engineering controls (such as barriers, enclosures and collection systems) and personal protective equipment are deployed and utilised.

PLANNING PHASE

Monitoring the work environment provides basic information on the extent and magnitude of the hazard potential and exposures of the workforce. The concept of air sampling and the use of air monitoring devices might appear to be simple, at first glance, but closer inspection reveals that there are many limitations and pitfalls for the unwary. Only by careful planning and preparation can meaningful results be generated.

The principal monitoring methods that are applied to workroom air are personal sampling and positional sampling. The former is used to assess the dose that an employee receives as a result of the concentration of the airborne substance and the duration of exposure. Sample collection equipment is placed as close as possible to the breathing zone of the wearer. It is self-contained and worn by the operator for all or part of the shift period. Positional sampling is used to establish the concentrations of a contaminant at a particular workroom location. It is useful in establishing the effectiveness of housekeeping initiatives and/or nearby engineering control measures.

Sometimes it can be used to give an indirect estimate of employee exposure or to indicate the presence of a potential hazard.

A characteristic of air contaminants in most occupational settings is that there is marked variability in concentration with respect to time and space. Air currents within the room, changes in work practices and variations in the emission rate of the contaminant are a few of the more important factors that give rise to this. The problem of establishing worker exposure is further complicated by the level of mobility associated with many jobs. If employee exposure is to be assessed, several considerations are involved:

(a)   Which employee or employees are to be sampled?

144

(b)   Where should the sampling device be located?

(c)   How many samples should be taken to define a representative exposure?

(d)   What sampling interval should be used?

(e)   During which periods of the work-day should the employee's exposure be sampled?

(f)   How many work-days during a year should be sampled and when?

These considerations provide the foundation for decisions on the types of instrumentation to be used and the methods of application.

ASSESSMENT PHASE

There are numerous sampling instruments available for measuring the concentrations of airborne substances. They are usually classified as:

(a)   direct-reading;

(b)   those that remove the substance from a measured volume of air for subsequent analysis; and

(c)   those that collect a measured volume of contaminated air for later analysis.

Direct-reading instruments are devices in which the sampling and analysis functions are carried out within the instrument, and the required information can be determined directly. Such instruments usually rely on physical phenomena, such as coulometry, electrical conductivity, thermal conductivity, combustion, chemiluminescence and aerosol photometry.

The second category of instruments are used in dealing with airborne particulates (such as dusts, fibres, fumes and mists), as well as gaseous samples. Particulates may be sampled to determine the total airborne concentration or the respirable fraction. As most samples are taken to determine the mass per unit volume of air, or to analyse the air for one or more specific substances, the procedure most commonly used is to pass the air through a suitable filter. For sampling of gaseous materials, a known volume of air is passed through an absorbing or adsorbing medium to remove the desired contaminant.

The third category of equipment involves introducing the sample of air into a collection device, such as an evacuated bottle, a flask or a plastic bag. Collected gases can be considered to be either "grab" samples or "integrated" samples, depending on the duration of sampling. After the sample of workplace air has been collected, the container is sealed and then sent to the laboratory for analysis. Sophisticated techniques, such as infrared spectrophotometry and gas chromatography, may then be used to determine concentrations of gaseous components.

The choice of a particular sampling instrument will depend upon factors such as the portability and ease of use of the device, its efficiency and reliability, type of analysis or information required, availability and past experience. There is no single, universal sampling instrument. All have limitations. It is important to know whether or not the particular instrument is specific for the contaminant to be determined, what other substances interfere with the test and the accuracy and sensitivity of the device.

At this point comment should be made on the aspect of calibration. Since the amount of contaminant collected will depend on the volume of air sampled, it is essential that the device operates at a known rate of airflow. Thus, the equipment must be calibrated against a standard airflow measuring device, both before and after use in the field. The exact rate of airflow must be recorded so that when it is multiplied by the sampling time, the total volume of air sampled or collected will be known. This is used in calculating the concentration of the contaminant to which the worker was exposed.

The final point in the assessment phase is to review the monitoring results in relation to recommended health standards and statutory limits for both short-term and long-term exposure. This, then, enables the magnitude of the problem to be established and paves the way for priority setting and planning for subsequent controls.

CONCLUSION

Virtually all work environments have potential or actual occupational hazards that need to be recognised, evaluated and controlled. Application of the strategies and techniques that have been reviewed in this paper will provide a means of quantitatively assessing the workplace environment and subsequently establishing conditions that will contribute to higher levels of safety, comfort and productivity.

REFERENCES:

1. Proctor, N.H. and Hughes, J.P. (1978). Chemical Hazards in the Workplace. J.B. Lippincott Co., Philadelphia.

2. Department of Employment and Industrial Relations (1984). Occupational Hygiene at Work — General Principles. Australian Government Printing Service, Canberra.

3. NIOSH (1977). Occupational Exposure Sampling Strategy Manual. U.S. Department of Health, Education and Welfare, Cincinnati, Ohio.

4. Harvey, B. et al (Eds) (1980). Handbook of Occupational Hygiene. Kluwer Publishing, Brentwood, Middlesex.

INDUSTRIAL HYGIENE AUDITS

by
Barry Chesson

INTRODUCTION

Industrial hygiene audits, whether they are of an internal (self-audit) or external style, are designed to examine the industrial hygiene practices within an organisation or a department with a view to establishing their effectiveness in preventing occupational illnesses and their conformance with corporate and Government standards and regulations. An audit should also establish whether programme documentation is likely to withstand third-party scrutiny.

Audit reports provide an indication of whether key industrial hygiene programme elements are present or absent. They are not designed to establish the degree to which elements of the programme have achieved the stated objectives, i.e. to reduce the incidence of noise-induced hearing loss within the organisation to a certain level. The presence of a particular element merely indicates that the organisation is capable of moving towards desirable preventive goals.

AUDIT TEAM

The auditing process might be carried out by an individual or by a team. Members might be specialists drawn from other company locations/operations, consultants or Government representatives carrying out their inspectorial functions.

If a team approach is to be employed, it is useful to include at least one site or departmental representative — this will assist with local support and awareness of the objectives and arrangements associated with the audit. Such a representative should be a person who is familiar with local processes, procedures, people and materials.

STEPS IN THE AUDITING PROCESS

(a)   Collate information on the site, department or function to be studied (i.e. written objectives, structure to address occupational health issues, previous audit reports, action plans, associated training activities, communication channels and control systems).

(b)   Use checklists, preliminary surveys, interview techniques, etc. to establish findings on the department or function.

(c)   Prepare interim recommendations and discuss these with the appropriate departmental or functional representatives.

(d)   Prepare final recommendations and confer with management.

(e)   Bind and file all audit documents.

AUDIT PROTOCOLS

Some sample audit work papers are appended.

Appendix I is an example of a data-gathering format that may be used during a preliminary survey of a site. Appendices II, III and IV are examples of sheets that have been applied successfully in evaluating practices within large industrial complexes.

Attention may be directed at the site overview (Appendix II) with elements such as industrial hygiene policy, structure/reporting relationships/manpower, analytical support facilities, communication/education and other functional considerations coming under scrutiny.

A third approach is to review activity associated with major topics, such as airborne contaminants, hearing conservation, vibration, sampling and analytical facilities, hazardous chemicals management, ionising radiation, non-ionising radiation, ergonomic hazards, exhaust ventilation, respiratory protection, potable water quality, illumination hazards, training/communication or enclosed space entry. Appendix III is an example of a checklist for one such topic — ionising radiation.

It may be desirable to examine the presence or absence of programme elements at departmental level (Appendix IV) — although this may involve a considerable amount of time and effort if each department on the site is to be covered.

In situations where there has been little previous audit activity, it may be necessary to combine all three approaches and conduct a single comprehensive investigation.

REFERENCES:

1.   Toca, F.M. (1981). Programme Evaluation: Industrial Hygiene. Am. Ind. Hyg. Assoc. J., 42; 213-217.

2.   Haas, B.H. (1982). Industrial Hygiene Audits. Am. Ind. Hyg. Assoc. J., 43: 867-873.

3.   Corn, M. and Lees, P.S.J. (1983). The Industrial Hygiene Audit: Purposes and Implementation. Am. Ind. Hyg. Assoc. J., 44: 135-141.

PRELIMINARY SURVEY DATA

1. BASIC DATA

Investigator ________________________________ Date ____________

Organisation Visited ____

Address ____

____

No. of Employees ______________________

Function of the Operation ______________________________________

____

Process Description ____

____

____

____

____

____

____

____

Raw Materials, By-products, End-
products ________________________________

____

____

____

Purpose of Visit ____

____

____

____

2.   **POTENTIAL HAZARDS**

Include Location Detail(s)

Check ( )

(a)   Physical   ____

____

(b)   Chemical   ____

____

(c)   Biological   ____

____

(d)   Ergonomic   ____

____

Other Details/Observations ________________________________

____

3.   **DEPARTMENT SURVEYED**

____

4.   **CONTROL MEASURES IN USE** (Ventilation, Personal Protective
Equipment)

____

____

____

____

____

5.   **SAMPLING DETAILS**

Agent(s) Selected for Survey _______________________________________

____

Sampling Location(s) ____

____

Type of Sample ____

____ No. of Samples ___________

Instrument(s) ____

____

Calibration Details ____

Sampling Rate _____________________________ Duration of Sampling __________

Other Details ____

____

____

6. **RESULTS**

____

____

____

____

____

____

7. **DISCUSSION**

____

____

____

____

____

____

8. **RECOMMENDATIONS**

____

____

____

____

____

INDUSTRIAL HYGIENE AUDIT
VOLUME I CHECKLIST: SITE OVERVIEW

A.   POLICY

1. Is there a readily available site policy statement that covers the industrial hygiene function? **YES/NO**

2. Is the policy consistent with corporate policy statements and policies of other operating sites? **YES/NO**

3. Is the policy complete? Does it clearly state the scope, responsibilities and authority of the programme? **YES/NO**

4. Is it understood and supported by management and employees? **YES/NO**

5. Does the policy carry the authority needed for implementation? **YES/NO**

B. STRUCTURE/REPORTING RELATIONSHIPS/MANPOWER

1. Is there a formal committee structure that allows an adequate interface between I.H. personnel and site management? YES/NO

2. Does the hygienist have adequate direct, informal access to site management? YES/NO

3. Does the hygienist servicing the site report to a senior manager? YES/NO

4. Are there adequate formal arrangements to facilitate liaison with hygienists at other sites? YES/NO

5. Are there suitable opportunities or arrangements in place to ensure that I.H. personnel interface with site medical, safety, engineering and environmental personnel? YES/NO

6. Are there formal structures in place to facilitate communication between the hygienist and safety representatives/union officials? YES/NO

7. Does the hygienist's current Position Description suitably describe the current nature of the job being performed. YES/NO

8. Are there sufficient manpower resources to cope with currently identified I.H. issues? YES/NO

9. Is the hygienist suitably trained? YES/NO

10. Are I.H. support personnel adequately trained for the task they are currently performing or expected to do in the future? YES/NO

11. Are support personnel at an appropriate job grade in the organisation? YES/NO

C. I.H. ANALYTICAL SUPPORT FACILITIES

1. Is the laboratory NATA approved?   YES/NO

2. Are I.H. analyses performed by trained chemists or technicians?   YES/NO

3. Does the laboratory have the necessary instruments and equipment required to perform analysis by approved methods?   YES/NO

4. Do the procedures and methods employed meet generally accepted standards?   YES/NO

5. Are calibration arrangements adequate?   YES/NO

6. Are the equipment maintenance/repair arrangements satisfactory?   YES/NO

D. **COMMUNICATION/EDUCATION**

1. Are results generated by industrial hygiene sampling communicated to the Departments concerned? YES/NO

2. Are participating employees provided with access to results? YES/NO

3. Are new employees provided with an I.H. orientation? YES/NO

4. Are site employees given regular presentations on I.H. topics? YES/NO

5. Is frequency more than twice per annum per employee? YES/NO

6. Are attendance records maintained in a satisfactory manner? YES/NO

7. Are dangerous areas suitably posted with warning signs? YES/NO

INDUSTRIAL HYGIENE AUDIT
VOLUME I CHECKLIST: SITE OVERVIEW

E.   OTHER FUNCTIONAL CONSIDERATIONS

1.   Is there a safety review procedure when new projects are contemplated?   YES/NO

2.   Is there an I.H. input to appropriate project safety review sessions?   YES/NO

3.   Does the hygienist deal with community concerns?   YES/NO

4.   Is there adequate liaison with public relations officials?   YES/NO

5.   Is the hygienist involved in accident investigations, as appropriate?   YES/NO

6.   Is there adequate I.H. inputs into written job procedures in plant operating areas?   YES/NO

7.   Does the hygienist have clearly established responsibilities in dealings with Mines Inspectors?   YES/NO

INDUSTRIAL HYGIENE AUDIT
VOLUME I CHECKLIST: MAJOR AGENTS REVIEW

A. IONISING RADIATION

1. Is there a written Radiation Control Programme? YES/NO

2. Have written plans been established to deal effectively with a radiation emergency? YES/NO

3. Are the responsibilities for managing site ionising radiation sources clearly established? YES/NO

4. Are minimum X-ray intensities and radioactive material quantities used to do the job? YES/NO

5. Have radiation devices been constructed so as to minimise the potential for radiation exposure? YES/NO

6. Have radiation devices been installed/positioned so as to minimise the potential for radiation exposure by maintenance workers? YES/NO

7. Are radiation devices located outside normally occupied areas and off normal access routes? YES/NO

8. Are sealed source radiation gauges positioned to be relatively protected in the event of a fire, mechanical damage, explosion or other major accidents? YES/NO

9. Are there routine checks of the status of the radiation sources? YES/NO

10. Is the measurement equipment satisfactory? YES/NO

11. Has each employee working with or around a radiation device been adequately trained?

  (a) within the Department responsible for maintenance of the devices? YES/NO

  (b) other employees? YES/NO

12.   Is there a sufficient number of qualified people to provide back-
        up to the nominated Radiation Safety Officer?   YES/NO

13.   Are warning signs satisfactory in terms of wording?   YES/NO

14.   Are warning signs satisfactory in terms of deployment?   YES/NO

15.   Are warning signs satisfactory in terms of condition?   YES/NO

16.   Are the following documents maintained on file?

        o   Up-to-date inventory of radiation sources.   YES/NO

        o   Initial radiation survey profiles.   YES/NO

        o   Job class exposure dose estimates.   YES/NO

        o   Periodic inspection checklists.   YES/NO

        o   Government inspection reports.   YES/NO

        o   Registration forms, as appropriate (i.e. X-ray, gamma
                sources, luminous discharge lighting).   YES/NO

        o   Written precautionary procedures for handling and ser-
                vicing radiation devices.   YES/NO

        o   Survey instrument calibration certificates.   YES/NO

        o   Purchase orders and miscellaneous receiving reports.   YES/NO

        o   Layout map indicating location of radiation source in the
                plant.   YES/NO

17.   Are the arrangements for renewal of licences satisfactory?   YES/NO

18.   Are the relevant legal requirements being met?   YES/NO

INDUSTRIAL HYGIENE AUDIT
VOLUME II CHECKLIST: DEPARTMENT ANALYSIS

Department:________________________

	Monthly	Fortnightly	Wages
No. of Personnel:________________ Days:	_______	_________	_____
Shift:	_______	_________	_____

Description of main work activities carried out within the Department:

____

____

____

____

____

____

Main physical agents encountered by the workforce:

____

____

____

Main chemical agents encountered by the workforce:

____

____

____

Audit party: _______________________________________

Date of Inspection:_________________________________

PERSONAL PROTECTIVE EQUIPMENT

by
Barry Chesson

GENERAL

Personal protective equipment (PPE) is appropriate in circumstances where there is no immediately feasible way to control the hazard by more suitable means (engineering or administrative controls), in emergency situations, or where it is employed as a temporary measure while more effective solutions are being devised.

The main shortcomings involved in the use of personal protective equipment are that it does nothing to eliminate or reduce the hazard; if the equipment fails for some reason, the worker may be exposed to a hazardous situation; and, finally, the equipment may be cumbersome and interfere with safe and effective performance of the task. Since there are limitations in terms of application, personal protective equipment should always be regarded as the last line of defence.

PPE include items of clothing, such as overalls, helmets, gloves, boots and aprons, and items of equipment such as earplugs, earmuffs, respirators, goggles, safety glasses, welders' masks and shields.

There are four key elements in a suitable PPE programme:

Selection

Clearly, the equipment must meet the basic criterion of providing adequate protection to cope with the particular workplace hazard against which it is being applied. It is necessary to take into account the nature of the hazard, the circumstances of the task to be performed (what, where, when, how, type considerations), the acceptable level of exposure for the hazard and the performance requirement for the device.

Fitting

Correct fit and comfort are essential if the calculated degree of protection is to be achieved.

For most items of PPE, a range of sizes is needed to accommodate the full range of shapes and dimensions of users. This may create administrative difficulties for the organisation, but is often the only method of ensuring that each user is supplied with equipment which correctly fits him.

Comfort is another important factor. If an item is uncomfortable, it is likely to be removed during at least part of the time when a hazard exists.

Maintenance and Storage

Poorly maintained equipment may result in serious health consequences. Some larger organisations use specialised contract or in-house services to collect, clean, repair and re-issue items of PPE. In some circumstances, attention to cleaning and maintenance of equipment, via recycling systems, can produce significant savings for the organisation.

Education and Training

It is important that PPE users be trained in the correct manner to use their equipment. Instructions should cover topics such as the need for the device, its design features, its applications and limitations.

RESPIRATORY PROTECTION

In dealing with airborne chemical agents, respiratory protection is the category of personal protective equipment of most importance. A hazardous or harmful atmosphere is one that is oxygen-deficient or contains a toxic particulate, vapour or gas in a concentration immediately or ultimately dangerous to life or health.

There are two fundamental alternatives in providing personal respiratory protection against such atmospheres:

(a)   use of air-purifying equipment; and

(b)   use of air-supplied equipment.

Selection of an appropriate device will depend on several important factors. The Australian Standard (AS 1715-1982) lists these as being:

(a)   adequacy of the warning given by the contaminant;

(b)   nature of the hazard — whether a particulate, gas or vapour, deficiency of oxygen, or a combination of these;

(c)   concentration of the contaminant;

(d)   acuteness of the hazard — if failure of the device for a short time can cause serious harm;

(e)   probable length of time during which the wearer will be in the contaminated atmosphere;

(f)   location of the contaminated atmosphere with respect to source of air suitable for breathing;

(g)   access to and the nature of the working environment in which the device is to be used;

(h)   expected activity of the wearer;

(i)   mobility of the wearer; and

(j)   whether the device is for regular use or for emergency or rescue purposes.

In the United States there has been considerable progress made over the last decade in the development and implementation of respiratory protection programmes. The Occupational Safety and Health Administration has produced a standard (29 CFR Part 1910. 134) which sets down the criteria for a minimally acceptable respiratory protection programme:

(a)   written standard procedures governing the selection and use of the respirators;

(b)   written records of instruction and training in the use of the respirator the employee receives;

(c)   respirators individually assigned, where possible;

(d)   regular cleaning and disinfecting;

(e)   respirator inspection;

(f)   respirator storage;

(g)   surveillance of work areas;

(h)   surveillance of employee exposure to contaminants;

(i)   regular inspection and evaluation of effectiveness of programme;

(j)   evaluation of individual usage; and

(k)   use of approved respirators.

Highly sophisticated programmes now exist within many industrial complexes. In setting up to meet the requirements of the standard, a number of organisations have chosen to work through the eight basic steps outlined on Appendix I.

Respiratory protection, like personal protective equipment in general, has a part to play is preventing adverse health effects arising from workplace activity. It is far more appropriate to control hazards by engineering or administrative means than to rely on protective equipment with all its limitations. Given that this last line of defence is necessary, comprehensive programmes of the type indicated will go a long way towards ensuring that work can be performed in a safe manner.

REFERENCES:

1. Standards Association of Australia (1982). Australian Standard 1715-1982: Selection, Use and Maintenance of Respiratory Protective Devices. S.A.A., North Sydney.

2. Standards Association of Australia (1986). Manual of Industrial Personal Protection: SAA HB9 – 1986. S.A.A. North Sydney.

3. Harvey, B. (Ed) (1980). Handbook of Occupational Hygiene. Kluwer Publishing Company, Brentford, Middlesex.

EIGHT STEPS FOR AN EFFECTIVE RESPIRATORY PROTECTION PROGRAMME

1. ADMINISTRATION

Put a person in charge of the whole programme. Provide workable links with others who can help, i.e. safety engineers, industrial hygienists and physicians.

2. DEFINING RESPIRATORY HAZARDS

Consider the possibility of oxygen-deficient atmospheres. Study all the contaminants that could be present in the breathing zone of workers. Determine the Threshold Limit Values (TLVs) of likely air contaminants.

3. HAZARD ASSESSMENT

Study the whole operations and locate every potential exposure hazard. Sample and test with suitable instruments during all conditions of the operations. Take samples in the breathing zone frequently enough to cover the range of typical exposures. Assess potential emergency conditions.

4. HAZARD CONTROL

Effective engineering controls should be used, wherever possible, to eliminate or reduce workers' exposure to respiratory hazards. Where this cannot be accomplished practically, proper protective devices must be provided for all exposed workers.

5. SELECTION OF RESPIRATORY PROTECTION

Use results from the hazard assessment step to select the devices which give the desired protection. Devices which have been tested and approved by the appropriate regulatory body should be used, whenever possible.

6. TRAINING

Before the need arises for workers to come into contact with potentially hazardous atmospheres, they should be thoroughly trained on the nature of the hazard and its potential harm, why it cannot be eliminated now, how the protective device works and its limits, and then be instructed on its use. This should be followed by field supervision.

7.   INSPECTION, MAINTENANCE AND REPAIR

Set up and follow a well-defined procedure for keeping protective devices as effective as the day they were purchased. This will vary widely according to plant size, nature of hazards and frequency of use. The important thing is a consistent, documented programme.

8.   MEDICAL SURVEILLANCE

In addition to a regular programme of pre-employment physical examinations, continuing medical monitoring will help determine the success of the respiratory protection plan.

HAZARDOUS MATERIALS MANAGEMENT
by
Barry Chesson

INTRODUCTION

In recent years there has been a significant increase in community awareness about the potential hazards posed by chemicals in the workplace. This has resulted in the current high level of activity by the Government, and union and employer groups.

This paper identifies the main elements of a site hazardous materials management programme, elaborates on mechanisms and tools of implementation and puts into perspective hazardous materials information as part of a wider health management system.

POLICY AND STRATEGY STATEMENTS

For an organisation intending to launch a hazardous materials management programme, a useful starting point is to produce a written statement of policy and a broad outline of strategy for giving effect to the policy statement. Model policy and strategy statements, as developed by the Chamber of Mines of Western Australia, are provided as Appendices I and II respectively.

SITE PROGRAMMES

Several of the larger industrial organisations in Western Australia have comprehensive hazardous materials management systems already in place. Such programmes cover raw materials, products, by-products and a wide range of consumable materials. The key components of a site hazardous materials management programme are as follows:

(a)   supply/industrial hygiene/medical involvement in vetting new chemicals and controlling access of the more toxic materials;

(b)   collation of supplier information, literature information and user experience;

(c)   development of manuals, summary sheets and supplementary systems;

(d)   development of systems to track deployment and consumption of chemicals;

(e)   development of procedures to cover manufacture, transport, storage, use and disposal of chemicals;

(f)   posting of warning signs;

(g)   labelling to conform with national standards;

(h)   institution of engineering, administrative and personal controls;

(i)   walk-through surveys;

(j)   industrial hygiene workplace measurements;

(k)   training and education of user groups; and

(l)   medical surveillance, where needed.

MATERIAL SAFETY DATA SHEETS (MSDSs)

MSDSs are designed to gather essential health and safety information from manufacturers/suppliers on the workplace materials that they provide. Many versions are available, but usually all contain provisions for gathering information on the identity and composition of the material, its physical and chemical characteristics, physical hazards, health hazards and emergency and first-aid procedures. An example of a typical MSDS is attached at Appendix III.

VETTING SYSTEMS

The MSDS system is primarily a means of providing information to those who work with chemical substances. The sheets also provide an opportunity for health professionals to examine data on the materials and determine any implications that they may have in terms of impact on health and safety at the work site. Such a review may result in a material being considered unacceptable for use on the site or, alternatively, it may only be used under certain conditions. For a large organisation, it is essential that the co-operation and involvement of the supply/purchasing function is obtained. Supply departments have day-to-day dealings with vendors and are able to apply commercial pressure if a supplier is reluctant to provide full and adequate information. Each time a new chemical is to be purchased, the same procedure should be followed. The attached chemical requisition flowchart (Appendix IV) is an example of the type of procedure to be followed.

HAZARD COMMUNICATION LAWS

Some countries now have laws which impose a general duty of care on employers with respect to protecting the health and safety of employees. Codes of practice and other devices are used to outline the mechanisms for coping with hazardous materials.

Other countries have more specific laws which oblige employers to instal comprehensive hazard communication programmes. In the United States, OSHA

Hazard Communication Standard (29 CFR Part 1910. 1200) was introduced recently with the dual purpose of ensuring that chemical hazards are evaluated and that hazard information is transmitted to employees. It required that organisations put in place arrangements in relation to written programmes, hazard evaluation, material safety data sheets, labelling and employee information and training.

MATERIALS INFORMATION AS PART OF A HEALTH MANAGEMENT PROGRAMME

A number of major corporations in Australia and overseas have developed, or are in the process of developing, computer-based health management systems. Such systems usually contain modules covering hazardous materials information, work area details, industrial hygiene data, medical records, job histories and employee header details. These may be used for the day-to-day management of occupational health of employees, or as a basis for epidemiological investigations.

The hazardous materials component is often the first module to be developed. It is usually considered a driving force for the health management system since, while it has no dependency on other modules for data, its data will be required for the operation of other sub-systems.

The module is based on an inventory of chemicals and material safety data sheet (MSDS) information. It is usually linked to a system, as described above, which provides for vetting of chemicals prior to entry to the site and periodic review of materials that have been in use for some time.

CONCLUSIONS

Chemical materials are an essential part of our day-to-day existence and contribute to the standard of living which we all enjoy. Some chemicals have properties such that the adverse health effects will not be experienced even after continuous day-after-day exposure. On the other hand, some may exert severe effects after relatively brief contact.

It is a wise precaution to treat all chemicals as potentially hazardous and, therefore, minimise exposure. Safeguards should be considered at all stages in the life cycle of the chemical: during manufacture, transport, storage, use and disposal. Management practices and initiatives, such as those outlined above, will reduce the potential for adverse effects to be experienced.

HAZARDOUS MATERIALS POLICY

This organisation shares the concern of the community regarding uncontrolled use of materials which are hazardous to the health or environment.

The organisation recognises that it has a responsibility to protect its employees, the environment and the community at large from any chemical hazards which are within its control. It is further recognised that this responsibility requires the establishment of procedures to ensure maintenance of health, safety and environmental standards during transport, use and disposal of hazardous materials. Such procedures must be designed to protect against short and long-term hazards to human health and the environment.

The organisation acknowledges that employees who handle, or may be affected by hazardous chemicals, have a right to be fully informed of the nature of those hazards. It is recognised that the provision of this information will assist individuals to minimise their exposure to chemical hazards.

The organisation believes that the responsibility for providing information on the potential hazards of chemicals and the procedures to be followed for safe transport, handling and storage rests firstly with the suppliers of those materials. The organisation recommends that member companies insist on provision of relevant information by suppliers prior to purchase of chemicals. At the same time, all employees have a responsibility to follow the health and safety procedures devised in the light of this information.

To give effect to this policy, the organisation will follow the appended strategy.

HAZARDOUS MATERIALS STRATEGY

1. Hazardous materials will be used only after taking into account the degree of risk involved and the operational and economic effects of substitution with less hazardous materials.

2. Where use of hazardous materials is essential, employee exposure and release to the environment of those materials or associated noxious effluents will be kept as low as reasonably practicable by engineering means. Relevant statutory limits will be regarded as the minimum standard of achievement in this respect.

3. As a last resort, where process substitution or engineering controls are not practicable or are inadequate for reduction of employee exposure, access to hazardous materials will be restricted to essential employees, who will be provided with appropriate, approved, personal protective equipment.

4. Where protective equipment is necessary, all affected employees will be thoroughly trained in its selection, fitting, use and, where appropriate, maintenance.

5. All personal protective equipment will be checked at appropriate intervals by responsible officers and repaired or replaced, where necessary.

6. All new processes and chemical products will be investigated for known potential hazards prior to implementation or purchase. Information will be obtained on the potential for and requirements for protection against acute and chronic hazards. In the first instance, this information will be sought from suppliers and will be supplemented from other sources, as necessary. Chemical products may not be purchased unless such information is available.

7. Hazardous materials may not be purchased unless they are packaged and labelled in accordance with the Australian Code for the Transport of Dangerous Goods by Road and Rail (1984) as a minimum specification.

8. Safe handling procedures, consistent with the best available knowledge, will be developed and implemented to cover transportation, storage, use and disposal of hazardous materials.

   These procedures will be based on supplier information, technical/research literature information, user experience and local conditions.

Emergency and first-aid procedures will also be developed and adequate warning signs and emergency equipment will be installed.

9. Employees who handle chemicals, or may be affected by them, have a right to be fully informed of the hazard potentials of those chemicals and the procedures for safe handling, minimisation of exposure and first-aid.

   Information on hazardous materials and safe handling procedures will be disseminated regularly to all employees via group and individual training, data sheets, manuals and other aids. The aim will be to ensure that safe handling procedures are both known and understood by all concerned.

10. Regular supervision and both occupational hygiene and environmental surveys will be undertaken in all areas where hazardous materials are handled.

11. Where appropriate, medical surveillance of employees will be undertaken.

MATERIALS SAFETY DATA SHEET

PLEASE ANSWER ALL QUESTIONS

SECTION 1: PRODUCT IDENTIFICATION

PRODUCT NAME	(1) REGULAR TELEPHONE NO.
SYNONYMS	(2) EMERGENCY TELEPHONE NO.
MANUFACTURER'S NAME AND ADDRESS	(1) (2)
DISTRIBUTOR'S NAME AND ADDRESS	(1) (2)
PRODUCT TYPE	CHEMICAL FAMILY

SECTION 2:   PRODUCT CONTENT

MATERIAL OR COMPONENT	PERCENTAGE RANGE (1)
MAJOR COMPONENTS	
MINOR COMPONENTS	

NOTE (1):   Where percentages are quoted, please specify whether by volume or by weight.

SECTION 3: USE INSTRUCTIONS

FOR WHAT PURPOSE IS THE PRODUCT USED?
WHAT ARE THE DIRECTIONS FOR USE?

SECTION 4:   PHYSICAL PROPERTIES

APPEARANCE AND ODOUR	pH
BOILING POINT ($^{\circ}$C)	Volatiles (% volume) (Method)
VAPOUR PRESSURE (KPa)	Evaporation (Butylacetate = 1)
VAPOUR DENSITY (Air = 1)	Melting Point ($^{\circ}$C)
SOLUBILITY IN WATER	Density (Kg/L)

SECTION 5: FIRE AND EXPLOSION DATA

FLASH POINT ($^{\circ}$C) METHOD USED	Flammability Limits in Air (% Volume)	Lower	Upper
FIRE EXTINGUISHING MEDIA			
SPECIAL FIRE-FIGHTING PROCEDURES			
UNUSUAL FIRE AND EXPLOSION HAZARDS			

SECTION 6: HEALTH HAZARD INFORMATION

TOXICITY DATA   (Please list and differentiate between effects of acute and chronic exposure.) EFFECT ON/OF:
EYES
SKIN
INHALATION
INGESTION

EMERGENCY AND FIRST-AID PROCEDURES

EYES
SKIN
INHALATION
INGESTION

SECTION 7: REACTIVITY DATA

STABILITY (PLEASE TICK APPROP. BOX)
    STABLE   ☐   CONDITIONS TO AVOID
    UNSTABLE ☐

HAZARDOUS DECOMPOSITION PRODUCTS

HAZARDOUS POLYMERIZATION
    WILL NOT OCCUR   ☐   CONDITIONS TO AVOID
    MAY OCCUR   ☐

SECTION 8: SPILL OR LEAK PROCEDURES

STEPS TO BE TAKEN IF MATERIAL IS LEAKED OR SPILLED

WASTE DISPOSAL METHOD

SECTION 9: SPECIAL PROTECTION INFORMATION

MATERIALS/SITUATIONS TO AVOID

VENTILATION REQUIREMENTS

RESPIRATORY PROTECTION

EYE PROTECTION

PROTECTIVE GLOVES

PROTECTIVE CLOTHING/OTHER

SECTION 10: SPECIAL PRECAUTIONS

HANDLING AND STORAGE PRECAUTIONS

OTHER PRECAUTIONS

NOTE: Please attach any further information available, e.g. product information sheets, container label and list of relevant references.

PREPARED BY	Title	Company	Date
APPROVED BY COMPANY MANAGER NAME		SIGNATURE	

CHEMICAL REQUISITION FLOWCHART

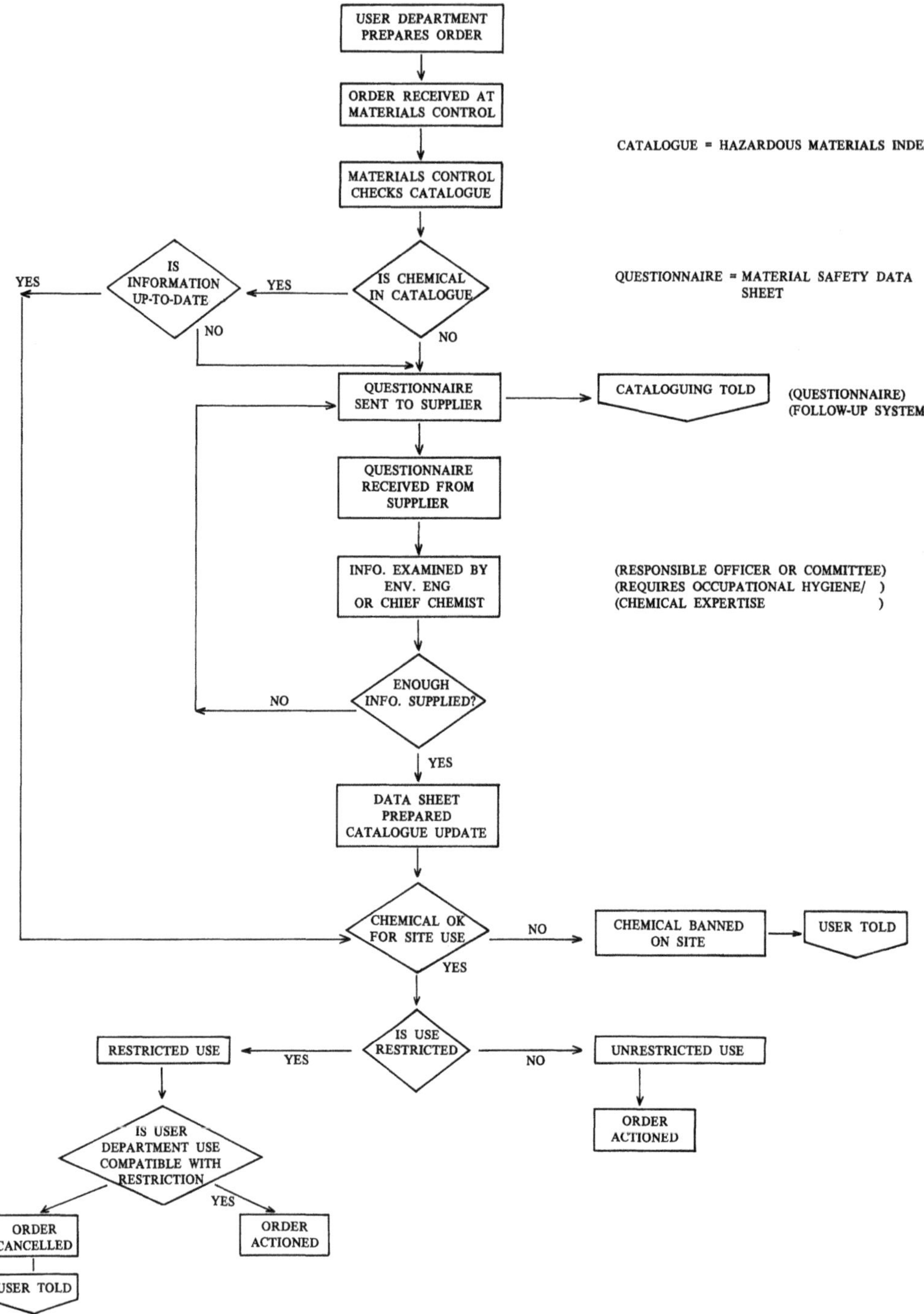

DISPOSAL OF HAZARDOUS WASTES

by
Barry Chesson

DEFINITION

There have been many attempts to establish what constitutes a hazardous waste. One such definition, provided by the Australian Environment Council, is as follows:

> *"A substance or mixture of substances which has no further economic use and which, if disposed of untreated to land, water or air, will be potentially harmful to man or his environment, by reason of its chemical, biological or physical properties".*

The four key properties that are linked to the above definition are toxicity (which includes considerations of mutagenicity and bio-accumulation), corrosivity, flammability and reactivity.

CLASSIFICATION

Wastes may be classified in several different ways. An approach used by the U.S. Congressional Budget Office, based on major constituent and physical state, is provided as Appendix I.

SOURCES

Industries with waste products likely to be defined as hazardous include:

(a)   mining;

(b)   textiles;

(c)   paper and printing;

(d)   chemicals, paint and petroleum;

(e)   refining, smelting and fabrication of metals; and

(f)   miscellaneous manufacturing industries, such as plastics, leather and rubber.

PROVISION FOR TREATMENT, STORAGE AND/OR DISPOSAL

Most waste materials are either stored, buried (on land or in the sea), burned or made to undergo a physico-chemical transformation. From time to time, this transformation may produce a marketable product.

Also, energy can be recovered from some burning processes. In this regard, some wastes have a comparatively high calorific value (i.e. solvents, oil sludges, etc.) which greatly improves the economics of treatment. Appendix II provides an outline of the main alternatives for treatment and storage/disposal.

ACCIDENTAL GENERATION AND DISSEMINATION

Chemical accidents, explosions and spills are major features of the hazardous chemicals issue. Accidents, including fire, can occur at any stage during the life cycle of a material. Many chemicals are relatively safe under normal conditions but may become hazardous when wet, heated or burnt.

Operations that routinely generate hazardous waste or have a potential to do so via leaks, spills or accidents should be governed by a hazardous waste management system. There should be a review made of the design and condition of the plant to determine the potential consequences of a release. All possible accidents and departures from the norm should be considered and then analysed in terms of the hazard and the corrective action which might be applied.

The main potential hazard in storage areas is catastrophic failure of a tank. Such an accident, on economic grounds alone, would justify close attention to tank design, maintenance and inspection. Containment of a large spill may be provided by diking or curbing: however, both systems need to be examined in the context of their operation when on a standby mode and in the event of a spill.

Loading, unloading and transfer operations are particularly accident prone, and so should be closely scrutinised.

The emphasis in addressing accidental generation of waste must be based equally on adequate, well-maintained equipment and on operational vigilance and supervision.

If an unacceptable release does occur, certain basic steps need to be taken to remedy the situtation:

(a)   source of the release needs to be located and capped;

(b)   physical containment of the spillage may be necessary, particularly if the release is in liquid form;

(c)   monitoring needs to continue in order to verify that the release has been brought under control;

(d)   vigilance may be warranted to ensure that there are no additional unacceptable releases;

(e)   physico-chemical treatment of the release (i.e. neutralisation) may be required, followed by clean-up and repair of any environmental damage.

EMERGENCY SERVICES

Accidental release of hazardous materials into the environment may necessitate the attendance of outside agencies. It is important that firemen and emergency service personnel have access to basic information on the nature of the chemical material that they are combatting. Many countries now require owners of premises or manufacturers to provide detailed information about the nature of their stored hazardous goods to local fire brigades and to the authority responsible for the maintenance of a hazardous chemicals response system. A standard code for the placarding of buildings containing dangerous goods is a desirable national goal. In the United Kingdom, the hazchem system has been a useful device in providing emergency service personnel with essential information on the action to be taken during the initial stages of a chemical incident. A simple numerical code indicates:

(a)   the correct fire-fighting medium to use;

(b)   the type of personal protective equipment required;

(c)   whether the chemical substance is likely to react violently;

(d)   whether to control spillage or run-off from the fire; and

(e)   whether there is need for public evacuation.

CONCLUSIONS

Chemical contamination and waste disposal are global issues, and are likely to remain so for some time to come. It is encouraging that many industrial organisations are now implementing comprehensive management programmes that contain both emergency response and preventive elements. In addition, Governments around the world have been active in recent years in introducing legislation to facilitate "cradle-to-grave" control over all aspects of hazardous management.

REFERENCES:

1.   Commonwealth of Australia (1982). Hazardous Chemicals: Report of the House of Representatives Standing Committee on Environment and Conservation. Australian Government Printing Service, Canberra.

2.   Yakovitz, H. (1985). Hazardous Waste Management: An Overview. In National Strategies for Managing Hazardous Waste. Commonwealth Department of Arts, Heritage and Environment, Melbourne, Australia.

3.   Worrall, M. (1985). Hazardous Waste Incineration for Australia. In National Strategies for Managing Hazardous Waste. Commonwealth Department of Arts, Heritage and Environment, Melbourne, Australia.

4.   Hackman, E.E. (1978). Toxic Organic Chemicals: Destruction and Waste Treatment. Noyes Data Corporation, Park Ridge, N.J.

5.   Congressional Budget Office (1985). Hazardous Waste Management: Recent Changes and Policy Alternatives. U.S. Government Printing Office, Washington.

A WASTE CLASSIFICATION SYSTEM BASED ON MAJOR CONSTITUTENT AND PHYSICAL STATE

Waste Type	*Examples*
LIQUIDS	
Waste oils	Spent crankcase oil, industrial lubricants.
Halogenated solvents	Spent trichloroethylene. Chloroform, Carbon tetrachloride
Non-halogenated solvents	Spent acetone, methyl ethyl ketone.
Other organic liquids	Aqueous organic solutions from cleaning or degreasing operations.
Metal-containing liquids	Metal-finishing solutions (acidic or alkaline).
Cyanide and metal liquids	Neutralised acid or basic washes with cyanide salts.
Polychlorinated biphenyls (PCBs)	Transformer fluids.
Non-metallic inorganic liquids	Acidic or basic solutions without metals.
SLUDGES	
Oily sludge	Tank bottoms, oil/water separation sludge.
Halogenated organic sludge	Halogenated still bottoms.
Non-halogenated organic sludge	Still bottoms without halogens.
Metal-containing sludge	Electroplating or chrome pigments, waste-water treatment sludges.
Cyanide and metal sludge	Metal heat treating sludges.
Non-metallic inorganic sludge	Sulphur sludge, lime sludge.
Dye and paint sludge	Heavy metal and solvent sludges.

Waste Type	Examples
SOLIDS	
Contaminated clay, soil, sand	Clay filters, spilled material.
Metallic dusts and shavings	Primary metal dusts and metal machinery wastes, emission control dusts from steel and lead industries.
Non-metallic inorganic dusts	Precipitator or baghouse wastes.
Halogenated organic solids	Polyvinyl chloride.
Non-halogenated organic solids	Polyethylene, cyclic intermediates.
MIXED	
Pesticides, herbicides	Pesticides, dioxins and other production wastes.
Explosives	TNT, wastewater treatment sludges from explosives production.
Miscellaneous wastes	Lab. waste chemicals, equipment containers, unspecified wastes.
Resins	Phenols, epoxy, polyester.

Source: Congressional Budget Office (United States).

TREATMENT AND DISPOSAL ALTERNATIVES

1.   Incineration (without energy recovery).

2.   Deep-well injection (injection of liquid discards into wells or salt domes).

3.   Surface impoundment (placement of liquid or sludge discards into pits, ponds or lagoons).

4.   Land farming or spreading of discards onto land.

5.   Land treatment. (bio-degradation of liquid or sludgy discards in soils).

6.   Especially engineered landfill (placement into lined discrete cells which are capped and isolated from one another and the environment).

7.   Ordinary landfill (placement of discards into a facility which receives normally inert refuse).

8.   Biological treatment ⎫
                       resulting in final
                       compounds or mixtures
                     ⎬ which are discarded to
9.   Physico-chemical   any of the processes in
      treatment   ⎭ this list

10.   Release into a marine body (discharge of treated or untreated liquids into municipal sewage treatment plants, rivers, streams, lakes or seas).

11.   Dumping into a marine body (off-loading of waste into the sea).

12.   Incineration at sea.

13.   Isolated permanent storage, e.g. drums stored in a specially engineered salt mine cavity.

14.   Incineration with energy recovery.

15.   Solvent reclamation/regeneration.

16.   Oil re-refining.

17.   Acid regeneration.

18.   Recycling/reclamation of non-ferrous metals.

19.   Recycling/reclamation of ferrous metals.

20.   Recycling/reclamation of a mix of ferrous and non-ferrous metals.

21.   Other recycling/reclamation.

22.   Long-term storage, e.g. more than 90 days.

ANNEXURE

SYLLABUS OF SEMINAR PROGRAMME

I. **Latest Concepts of Accident Prevention**

   A. Accident process and accident prevention strategies

   -   accident case studies
   -   process of accident sequence
   -   accident prevention strategies
   -   post-accident strategies, aimed at minimising accident consequences

II. **Modern Approach to Chemical Hazards Control**

   A. Identification of chemical hazards in industrial processes

   B. Management of chemical accident prevention systems

   -   plant integrity:
       (a) design standards
       (b) protective devices
       (c) maintenance
       (d) inspection and testing
   -   operating procedures

   C. Emergency Operations

   -   consequence criteria
   -   development controls
   -   emergency planning

   D. Training activities, exchange of information

   E. Low-cost methods to control chemical hazards

III. **Safety and Health Aspects of Storage and Handling of Hazardous Chemicals**

    A.   Developing Inspection Skills for the Prevention of Chemical Accidents

        1.   Reactive chemicals and the lines of defence. Safety aspects of plant layout

        2.   Safe storage and handling of dangerous materials
- flammable materials
- explosives
- materials liable to dust explosion
- evaluation and control of hazards due to self-heating
- materials sensitive to water and water vapour
- materials sensitive to acids and acid fumes
- corrosive materials.

    B.   Developing Inspection Skills for the Identification and Control of Chemical Health Hazards

        1.   Health hazards of nominated industries in Hong Kong

        2.   Principles and practice of industrial hygiene:
- legislation approaches to occupational health
- exposure standards for chemical substances
    - (a) hygiene limits
    - (b) synergistic and antagonistic effects
    - (c) hypersusceptibility
- chemical hazards, overview of recognition, evaluation and control aspects
- hazardous material management, policies and procedures
- sampling strategies for workplace contaminants and toxic releases
- industrial hygiene audits
- personal protection
- specific hazards related to industries in Hong Kong

IV.   **Techniques for Identification and Quantification of Chemical Hazards**

    A.   Assessment of the consequences of fire and explosion

    B.   Assessment of the consequences of toxic releases

    C.   Hazard analysis

    D.   Hazard and operability studies

    E.   Dow overall fire and explosion index technique

    F.   Techniques of inspections

V.   **Management of People in Chemical Accidents**

    A.   Disaster communication principles

    B.   Mass casualty procedures

        — command, control, communication

        — Triage

        — medical equipment

        — preventive medicine/public health.

    C.   Immediate medical consequences (detail contents depending on further discussions), e.g.:

        — blast (ear)

        — toxic gases

        — fire (eye, skin)

        — aircraft (fire)

        — radiation

        — natural phenomen associated

        — toxic gas/fumes/dust affecting food-chains.

    D.   Disposal of large-scale wastes/toxics following a chemical accident.

9 789221 062974